Spring Boot 从零开始学

/视频教学版/

郭浩然 编著

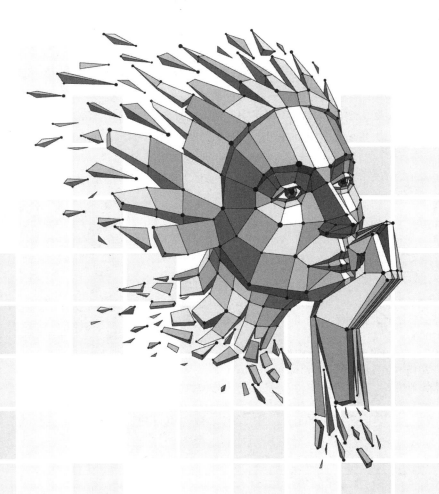

清华大学出版社
北京

内 容 简 介

Spring 是 Java 开发必不可少的框架，是一个庞大的生态系统，Spring Boot 正是在 Spring 这一片肥沃的土壤中生长出来的快速应用开发框架。本书从 Spring Boot 的工程化组件入手，采用一步一示例的方式引导读者入门，并通过两个完整案例帮助读者学会开发 Spring Boot 项目。本书配套源码、PPT 课件、教学视频、思维导图、开发环境与答疑服务。

本书共分 12 章。第 1～3 章介绍构建 Spring Boot 项目的步骤和 Spring Boot Web 开发的基础；第 4～9 章介绍 Spring Boot 的多种框架或技术，比如 JPA、MyBatis、Redis、Elasticsearch、日志和安全监控等；第 10～11 章分别通过客户管理系统和微博系统两个简单的项目，以巩固读者对 Spring Boot 基础知识的学习；第 12 章介绍 Spring Boot 的测试、打包和项目部署功能。

本书从实用的角度出发，结合项目示例，在充分实践的前提下尽量拓展知识广度、减少知识盲点，非常适合时间紧张却需要快速上手 Spring Boot 的初学者，也可作为高等院校、中职学校和培训机构计算机软件开发专业的教材。

本书封面贴有清华大学出版社防伪标签，无标签者不得销售。
版权所有，侵权必究。举报：010-62782989，beiqinquan@tup.tsinghua.edu.cn。

图书在版编目（CIP）数据

Spring Boot 从零开始学：视频教学版/郭浩然编著. —北京：清华大学出版社，2022.1（2024.8重印）
ISBN 978-7-302-59722-3

Ⅰ. ①S… Ⅱ. ①郭… Ⅲ. ①JAVA 语言－程序设计 Ⅳ. ①TP312.8

中国版本图书馆 CIP 数据核字（2021）第 277509 号

责任编辑：夏毓彦
封面设计：王　翔
责任校对：闫秀华
责任印制：刘　菲

出版发行：清华大学出版社
网　　址：https://www.tup.com.cn，https://www.wqxuetang.com
地　　址：北京清华大学学研大厦 A 座　　　邮　编：100084
社 总 机：010-83470000　　　　　　　　　　邮　购：010-62786544
投稿与读者服务：010-62776969，c-service@tup.tsinghua.edu.cn
质量反馈：010-62772015，zhiliang@tup.tsinghua.edu.cn

印 装 者：北京同文印刷有限责任公司
经　　销：全国新华书店
开　　本：190mm×260mm　　　印　张：20　　　字　数：540 千字
版　　次：2022 年 2 月第 1 版　　　　　　　印　次：2024 年 8 月第 4 次印刷
定　　价：75.00 元

产品编号：090167-02

前　　言

读懂本书

Spring Boot 能做什么？

Spring Boot 定义了大量的自动配置，能够根据环境、条件自动配置项目的组件，使用 Spring Boot 来开发可以大量减少我们的配置工作，提高开发效率。

Spring Boot 和 Spring 有什么区别？

Spring Boot 作为 Spring 体系的一部分，使用 Spring Boot 能够很方便地将 Spring 体系的其他模块整合到项目中。Spring Boot 是一个能够帮助我们整合、配置其他框架的框架。

Spring Boot 的学习和使用容易吗？

Spring Boot 在方便开发的同时也增加了学习成本，如果自学，则需要先掌握相关框架的知识。本书在介绍 Spring Boot 时考虑到了这个问题，所以在介绍 Spring Boot 时适当地介绍了相关技术的知识，争取做到不因为相关技术而影响读者对 Spring Boot 的学习和掌握。

本书真的适合你吗？

本书适合希望能快速上手 Spring Boot 的初学者，如果你恰巧想要快速地学会如何使用 Spring Boot，那么本书适合你！

本书涉及的技术或框架

Spring Boot	MyBatis	Log4j2	JavaScript
Spring MVC	Redis	SLF4J	jQuery
Spring Security	Elasticsearch	Thymleaf	Chrome 控制台
Spring Data	MySQL	HTTP	
Spring Data JPA	Logback	HTML	

本书涉及的示例和案例

修改 Web 服务端口号	自定义日志组
修改 Spring 的 Banner	指定日志文件名
使用 JpaRepository	禁用所有 Endpoint 的 JMX 实现
数据分页查询	设置 Endpoints 的端口使用 8081
使用@Query	通过 Metrics Endpoint 查看 JVM Metrics
自定义 Repository 接口	使用 Spring Data JPA 间接依赖 spring-jdbc
在 Redis 中保存键值对	使用 MyBatis 间接依赖 spring-jdbc
不指定序列化器时生成前缀字符	客户管理 Web 系统

使用 RedisCallback 查询数据	个人博客系统
使用 SessionCallback 执行多条命令	通过命令行启动项目
给非数字值使用 INCR 命令	验证配置的覆盖顺序
使用 GETBIT 命令	使用 JsonTest 测试 JSON 日期格式配置
使用 WATCH 命令实现对数据的自增操作	查看 WAR 包的目录结构
使用 SpEL 选取形参或属性	直接运行 Spring Boot 项目的 WAR 包
使用 Log4j 2	

本书的特点

（1）结合本书学习 Spring Boot 可以节省大量阅读官方文档或者网络博客的时间，能够轻松上手工程代码，在示例和项目实战中快速建立起对 Spring Boot 的认识。

（2）本书内容丰富，在介绍 Spring Boot 时，对涉及的相关技术做了适当的补充，不会因为对关联技术认识的缺失而阻碍对 Spring Boot 的学习。

（3）本书在撰写时结合了 Spring Boot 新版本和官方文档，在技术更新迭代迅速的如今，最大限度地避免了学完即过时的尴尬境地。

（4）对初学者友好。书中的示例和实战项目提供了比较详细的步骤和代码，可以一步一步跟着实践操作。另外，为了更专注地介绍 Spring Boot，在案例实战中的技术选型，与 Spring Boot 无关的技术都是选用简单或基本的技术，避免为初学者增加额外的学习负担。

示例源码、PPT 课件、教学视频等资源下载

本书配套的示例源码、PPT 课件、教学视频、思维导图、开发环境，需要使用微信扫描右边的二维码下载，也可按页面提示把链接转发到自己的邮箱中下载。如果有疑问，请联系 booksaga@163.com，邮件主题写"Spring Boot 从零开始学"。

面向的读者

- Java 开发工程师。
- Java Web 开发工程师。
- Web 应用开发人员。
- 没有使用过 Spring Boot 但希望学习 Spring Boot 的工程师。
- 使用过 Spring Boot 并希望深入掌握的开发人员。
- 了解 Java 基本语法，想要学习 Java Web 开发的初学者。

作　者
2022 年 1 月

目录

第 1 章　从零起步搭建 Spring Boot 开发
　　　环境 ································· 1
　1.1　Spring Boot 为什么流行起来 ············ 1
　　　1.1.1　Spring Boot 的优点 ············· 1
　　　1.1.2　Spring Boot 的时代背景 ·········· 2
　1.2　搭建开发环境 ··························· 2
　　　1.2.1　使用 Spring 官方提供的初始化
　　　　　　工具 ··························· 2
　　　1.2.2　搭建 Eclipse 开发环境 ··········· 5
　　　1.2.3　搭建 IntelliJ IDEA 开发环境 ····· 9
　1.3　Spring Boot 的依赖管理和自动配置 ····· 11
　　　1.3.1　依赖管理 ······················ 12
　　　1.3.2　自动配置 ······················ 13
　1.4　实战——Spring Boot 版本的
　　　Hello World ·························· 15

第 2 章　工程项目使用 Spring Boot 的
　　　步骤 ································ 18
　2.1　构建项目 ····························· 18
　　　2.1.1　构建工具 ······················ 18
　　　2.1.2　Starter、JAR 与依赖 ··········· 19
　　　2.1.3　再说依赖管理 ·················· 22
　2.2　组织代码 ····························· 23
　　　2.2.1　不建议使用 default package ····· 23
　　　2.2.2　放置应用的 main 类 ············ 24
　2.3　配置类 ······························· 24
　　　2.3.1　导入其他配置类 ················ 24
　　　2.3.2　导入 XML 配置 ················ 25
　2.4　再说自动配置 ························· 25
　　　2.4.1　用户配置替换自动配置 ·········· 25
　　　2.4.2　指定禁用生效的自动配置类 ······ 26
　2.5　Spring Bean 与依赖注入 ··············· 26
　2.6　使用@SpringBootApplication 注解 ····· 28
　2.7　运行程序 ····························· 28
　　　2.7.1　在 IDE 中运行 ················· 29
　　　2.7.2　打成 JAR 包运行 ··············· 29

　　　2.7.3　使用 Maven 插件运行 ··········· 30
　　　2.7.4　使用 Gradle 插件运行 ·········· 30
　　　2.7.5　热部署 ························ 30
　2.8　开发者工具 ··························· 30
　　　2.8.1　默认配置 ······················ 31
　　　2.8.2　自动重启 ······················ 31
　　　2.8.3　使用 LiveReload 自动刷新 ······ 34
　　　2.8.4　全局设置 ······················ 34
　2.9　打包应用到生产环境 ··················· 34
　2.10　实战——使用 Maven 创建完整的
　　　　工程项目 ···························· 34

第 3 章　使用 Spring Boot 进行 Web 开发 ··· 39
　3.1　模板引擎 ····························· 39
　3.2　使用 Thymeleaf 开发示例 ·············· 40
　3.3　上传文件 ····························· 43
　　　3.3.1　POM 文件配置 ················· 43
　　　3.3.2　参数设置 ······················ 43
　　　3.3.3　编写前端页面 ·················· 44
　　　3.3.4　编写处理上传请求的
　　　　　　Controller 类 ·················· 44
　　　3.3.5　从浏览器上传文件 ·············· 45
　3.4　使用定时任务 ························· 46
　　　3.4.1　POM 包配置 ··················· 46
　　　3.4.2　对自动配置参数的说明 ·········· 46
　　　3.4.3　编写定时任务代码 ·············· 47
　　　3.4.4　测试定时任务执行 ·············· 48
　3.5　发送邮件 ····························· 49
　　　3.5.1　POM 包配置 ··················· 49
　　　3.5.2　在 application.properties 中
　　　　　　添加邮箱配置 ··················· 49
　　　3.5.3　编写邮件 Service 类对框架
　　　　　　再封装 ························· 50
　　　3.5.4　编写测试类进行测试 ············ 51
　3.6　使用 Shiro ···························· 52
　　　3.6.1　基本配置 ······················ 52

	3.6.2 编写业务逻辑代码和页面·······53	5.3.3	Redis 发布订阅··············126
	3.6.3 在代码中引入 Shiro ···········55	5.4	使用 Spring 缓存注解操作 Redis·······127
	3.6.4 测试用户认证和权限管理的		5.4.1 启用缓存和配置缓存管理器···127
	效果························58		5.4.2 使用缓存注解开发·········128
3.7	实战——开发一个简单的 Restful API		5.4.3 类实例方法类内部调用时的
	网关······························59		失效问题···················129
第 4 章	**使用 Spring Boot 进行数据库**		5.4.4 缓存脏数据说明···········130
	开发························64	5.5	实战——用 Redis 改版商品信息管理
4.1	配置数据源·························64		系统 V2.0·······················130
	4.1.1 启动默认数据源············64		5.5.1 引入 Redis 的依赖并配置 Redis
	4.1.2 配置自定义数据源··········65		服务地址和启用缓存·······131
4.2	使用 JdbcTemplate 操作数据库·······66		5.5.2 添加@Cacheable 和@CacheEvict
4.3	使用 Spring Data JPA（Hibernate）		注解······················131
	操作数据·························67		5.5.3 运行程序测试缓存效果·····133
	4.3.1 基础知识·················67	**第 6 章**	**Spring Boot 整合 Elasticsearch····135**
	4.3.2 依赖管理和配置信息········68	6.1	Elasticsearch 的使用场景和相关
	4.3.3 使用 Spring Data JPA 进行		技术·····························135
	开发······················69	6.2	spring-data-elasticsearch 支持的
4.4	整合 MyBatis 框架·················85		Elasticsearch Client ··············136
	4.4.1 MyBatis 简介··············85		6.2.1 Elasticsearch 的 Client ······136
	4.4.2 MyBatis 的配置············85		6.2.2 创建 RestHighLevelClient····136
	4.4.3 Spring Boot 整合 MyBatis····88	6.3	使用 operations 相关 API 操作
	4.4.4 MyBatis 的其他配置········89		Elasticsearch ·····················137
4.5	实战——商品信息管理小系统·······90		6.3.1 4 个 Operations 接口·······137
第 5 章	**Spring Boot 与 Redis ···········103**		6.3.2 搜索结果类型············142
5.1	使用 spring-data-redis 操作 Redis······103		6.3.3 查询条件的封装··········143
	5.1.1 Spring Data Redis 项目的	6.4	Repository 的使用·················145
	设计······················103		6.4.1 使用注解管理索引实体类···145
	5.1.2 RedisTemplate 与数据操作类的		6.4.2 查询方法的定义··········147
	使用······················106		6.4.3 使用@Query 注解定义查询···148
	5.1.3 RedisCallback、SessionCallback	6.5	在 Spring Boot 中配置
	接口和 Redis 事务的使用····108		spring-data-elasticsearch ·········149
5.2	在 Spring Boot 中配置和使用 Redis···109	**第 7 章**	**Spring Boot 的日志管理··········150**
	5.2.1 通过 Starter 引入 Redis 相关依赖	7.1	常用的日志框架··················150
	并配置 Redis··············110		7.1.1 日志实现···············150
	5.2.2 Redis 数据类型及操作 API····112		7.1.2 日志门面···············151
5.3	Redis 的一些特殊用法·············125	7.2	Spring Boot 支持的日志配置·······152
	5.3.1 Redis 事务···············125		7.2.1 Spring Boot 默认支持的日志
	5.3.2 Redis Pipelined 和 Lua 脚本···126		框架······················152

7.2.2 自定义日志配置·················153
7.2.3 日志框架的配置文件···········157
7.2.4 配置项汇总·····················157
7.2.5 日志级别·······················159
7.2.6 日志格式和内容················160
7.2.7 输出到控制台··················162
7.2.8 日志组··························163
7.3 输出到日志文件的配置·················164
7.3.1 配置输出到日志文件···········164
7.3.2 日志滚动配置···················165
7.4 配置文件扩展···························166
7.4.1 定义 Profile 的个性配置······166
7.4.2 引入 Spring Environment Property·······················167

第 8 章 Spring Boot 的安全与监控········168
8.1 安全控制（使用 Spring Security）·····168
8.1.1 Spring Security 的开启和配置······························168
8.1.2 使用 Spring Security·············171
8.2 使用 Actuator 监控应用···············178
8.2.1 开启 Actuator····················178
8.2.2 默认配置·························179
8.2.3 Actuator 的安全控制···········182
8.2.4 Health Endpoint 的使用·······186
8.2.5 Metrics Endpoint················190
8.2.6 自定义 Endpoint················195

第 9 章 Spring Boot 数据访问···············199
9.1 自动配置默认数据源···················199
9.2 自定义一个或多个数据源·············204
9.2.1 在使用默认数据源实例的基础上自定义配置·············204
9.2.2 配置多个数据源·················205
9.3 Spring Data JPA 与数据源绑定·······207
9.4 数据库的初始化·························211
9.4.1 基于 SQL 脚本初始化数据库····························212
9.4.2 使用 JPA 和 Hibernate 时初始化数据库·····················213

第 10 章 项目实战 1——客户管理 Web 系统·························214
10.1 梳理业务需求·························214
10.2 技术实现设计·························215
10.3 构建项目································216
10.3.1 使用 Spring Initializr 构建项目·························216
10.3.2 配置数据库·····················220
10.4 创建数据库表·························221
10.5 开发客户信息模块····················222
10.5.1 开发系统首页··················222
10.5.2 开发添加客户页面和接口·····223
10.5.3 开发客户列表页面和接口·····228
10.5.4 开发编辑客户信息页面和接口····························232
10.6 开发交易信息模块····················236
10.6.1 在系统首页增加交易信息导航·························236
10.6.2 开发"创建交易"页面和接口····························236
10.6.3 开发"交易列表"页面······242
10.6.4 开发"编辑交易"页面······243
10.6.5 开发标注发货状态功能·······245

第 11 章 项目实战 2——个人博客·········249
11.1 梳理业务需求·························249
11.2 技术实现设计·························250
11.2.1 博客模块·······················250
11.2.2 用户模块·······················251
11.2.3 喜欢、取消喜欢博客功能····252
11.3 构建项目································253
11.4 创建数据实体类······················255
11.5 开发博客模块·························258
11.5.1 开发发布博客接口和页面····259
11.5.2 开发博客列表接口和页面····262
11.5.3 开发博客详情接口和页面····264
11.5.4 实现浏览次数计数功能······266
11.6 开发用户模块·························267
11.6.1 开发登录相关接口············267
11.6.2 完成登录页面··················273

11.6.3 测试用户登录功能 ············ 276
11.6.4 在博客列表页面增加当前用户
        的显示 ··························· 277
11.6.5 个人主页页面 ················ 280
11.7 实现喜欢/取消喜欢博客功能 ········ 283
11.7.1 开发"喜欢博客"接口 ····· 283
11.7.2 开发"取消喜欢博客"
        接口 ······························· 284
11.7.3 修改博客详情页面接口，
        返回当前用户是否已喜欢 ··· 285
11.7.4 修改博客详情页面，增加
        喜欢/取消喜欢按钮 ········· 286
11.7.5 页面测试 ······················· 287
11.8 配置 Spring Security 访问规则 ········ 289
11.8.1 创建管理员用户 ············· 289
11.8.2 配置接口的访问权限 ········ 290

11.8.3 配置仅管理员用户可以发布
        博客 ······························· 291
11.8.4 测试发布博客权限管理 ····· 293

第 12 章 Spring Boot 项目的测试和部署 ··························· 295

12.1 配置的切换 ······························· 295
12.1.1 在项目启动时指定外部
        配置文件 ························ 295
12.1.2 Spring Profile 的使用 ········ 299
12.2 Spring Boot 的测试功能 ··············· 301
12.2.1 构建测试类 ···················· 301
12.2.2 测试的自动配置 ············· 305
12.3 打包和部署 ······························· 308
12.3.1 打包（JAR 和 WAR）······· 308
12.3.2 运行项目 ······················· 310

第 1 章

从零起步搭建Spring Boot开发环境

简而言之，Spring Boot就是框架的框架，其对框架的自动配置为开发带来了非常大的便利。在搭建Spring项目时，常要进行复制、粘贴、修改烦琐配置文件等重复性工作，而使用Spring Boot就能解决这些重复性配置工作。Spring Boot正是为简化开发而生的。本章将通过搭建Hello World程序来帮助我们快速了解和上手Spring Boot。

本章主要涉及的知识点有：

- Spring Boot发展背景。
- Spring Boot的优缺点。
- 使用Spring Initializr快速构建项目。
- 使用Eclipse、IDEA构建开发环境。
- Spring Boot项目文件结构和自动配置。

1.1 Spring Boot为什么流行起来

Spring框架为简化项目开发而生，但是随着发展，Spring体系变得十分庞大，对框架之间的依赖管理和配置变成了项目开发中的新痛点。Spring Boot诞生的基础是对于框架之间的整合提供非常好的支持，但仅此还不能支撑Spring Boot被广泛地应用。本节将介绍Spring Boot都有哪些优点。

1.1.1 Spring Boot 的优点

1. 能够轻松创建一个独立运行的 Spring 程序

Spring Boot默认内置了Tomcat作为Web服务器，可选Jetty或Undertow，从而可以将Web应用程序打包成可执行的JAR文件或者是WAR文件，通过java -jar命令来运行应用程序。

2．提供 Starter 来简化配置

Starter是Spring Boot中定义的一系列依赖管理"快捷方式"，通过Starter可以完成必要的依赖管理、框架整合与配置工作。Spring官方以及第三方框架提供的Stater覆盖了常用的配置，我们在开发时仅需引入响应框架的Starter，便可直接开始使用相应的框架，做到"开箱即用"。

3．提供准生产级别的功能

Spring Boot提供了对程序运行的监控功能和配置功能。监控功能如程序的运行状况和程序的健康检查等。配置功能是在程序运行时对程序参数的查看、修改等。这些功能并不难，也有独立框架实现了这些功能，但是直接使用Spring Boot可以减少烦琐的工作和不必要的配置麻烦。

4．非侵入式

使用Spring Boot不会生成项目外的代码，并且无须引入XML配置文件。

1.1.2　Spring Boot 的时代背景

Spring Boot的流行与微服务、分布式和云计算这样的时代背景密不可分。传统的架构显得笨重、烦琐。微服务、分布式的发展产生了新的需求，如应用之间的通信、部署以及上线后的管理等。更高级的需求也有很多，比如服务之间的轻量交互、应用独立部署、服务发现、负载均衡、链路追踪等。

应对这些新的需求，Spring Boot以及继承Spring Boot风格的Spring Cloud提供了完整的实现策略，很好地满足了这些新的需求，而这些都进一步促进了Spring Boot被广泛应用。

1.2　搭建开发环境

搭建Spring Boot的开发环境就是构建确定的目录结构、添加必要的依赖。这些都是重复性工作，Spring官方已经为我们提供了快速构建开发环境的工具。本节将学习在Eclipse和IDEA中快速搭建开发环境。Eclipse和IDEA是开发Java程序常用的IDE，读者二选一即可。本书截图和示例采用IDEA。

1.2.1　使用 Spring 官方提供的初始化工具

Spring提供了Initializr工具，这是一个可视化界面，用来构建项目初始环境，如图1.1所示。Initializr工具分为6个模块，下面一一说明。

1．选择项目构建工具

这里有Maven和Gradle可选，Maven是比较经典的构建工具，不过Gradle变得越来越流行。两种工具管理的项目文件目录结构不同，配置文件也不同，在开发时可根据开发习惯和项目需要来选择，本书所有示例均以Maven为准。

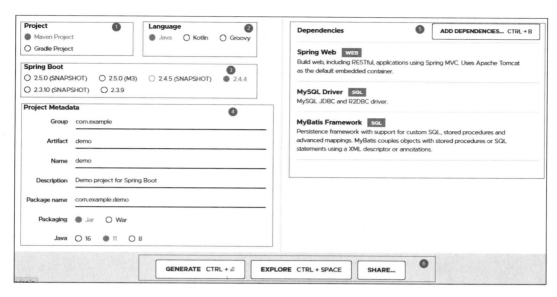

图 1.1　Spring Initializr

2．选择编程语言

这里提供了 Java、Kotlin 和 Groovy，根据实际开发需要选择即可。Web 后端开发一般选择 Java。

3．选择 Spring Boot 的版本

此时 GA 版本为 2.4.4，通常不会选择快照版（SNAPSHOT）。读者可在网站 https://spring.io/projects/spring-boot#learn 上查看当前版本，如图 1.2 所示。

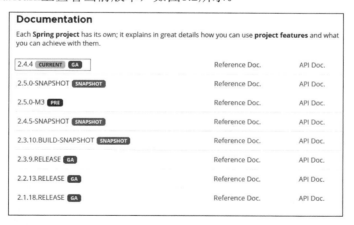

图 1.2　Spring Boot 版本

4．项目元信息

这里我们更多关注的是 Group 和 Artifact，它们分别对应 groupId 和 artifactId。Package Name 用来指定程序启动类所在的包名，也是代码最外层的包名，默认使用 Group.Artifact，也可以修改。还可以指定打包方式和 JDK 版本。打包方式有 JAR 和 WAR，分别是打成 JAR 包和 WAR 包。

5．选择依赖

单击按钮ADD DEPENDENCIES，打开如图1.3所示的窗口，为项目配置框架依赖。选中的依赖会写入配置文件中，待项目在开发环境中构建时自动下载JAR包。在图1.1中，已选择Spring Web、MySQL Driver和MyBatis这3个框架。

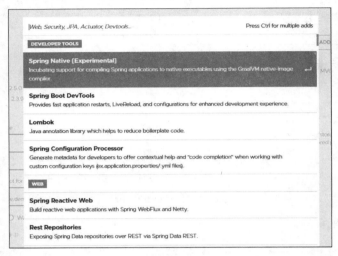

图 1.3　浏览器发出的请求 Body 内容

6．工具栏

工具栏有3个按钮：单击GENERATE按钮，可将初始化的项目目录和配置文件打包成artifactId.zip文件并下载；单击EXPLORE按钮，可预览项目目录和配置文件，如图1.4所示；单击SHARE按钮，生成可供分享的链接，如图1.5所示。我们单击GENERATE按钮，将demo.zip保存到本地。

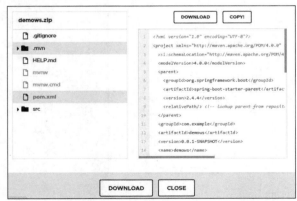
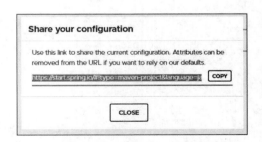

图 1.4　预览项目　　　　　　　　图 1.5　Spring Initializr 的分享功能

7．查看文件结构

压缩包中只包含配置文件和项目目录结构，大小仅有56KB。将demo.zip解压，demo目录如图1.6所示。

其中，src是源代码和配置文件的目录，pom.xml是项目的Maven配置目录，这是我们所关心的。其余的.mvn、HELP.md、mvnw和mvnw.cmd是提供给Maven程序的目录，.gitgnore用来配置git忽略策略的文件，这些我们通常不用关心。

使用命令"tree /f src"查看src目录结构，如图1.7所示。

图 1.6　Initializr 生成的目录结构

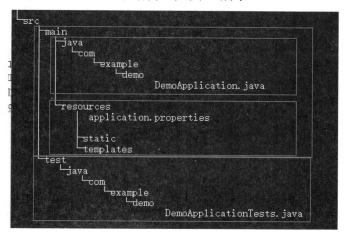

图 1.7　src 目录结构

src目录有两个子目录，即main和test。test目录用来存放测试代码，DemoApplicationTest.java是Spring生成的一个Test类。通过Maven配置可以在打包时忽略test目录下的代码。

main目录也有两个子目录，即java和resources。顾名思义，java目录下是JAR包和代码文件，其中的DemoApplication.java是Spring生成的程序启动类。

resources目录下存放配置文件和静态资源。此处由于在选择依赖时勾选了Spring Web框架，Spring Initializr除了生成配置文件application.properties外，还额外生成了static和templates目录，分别用来存放静态资源（如图片、CSS文件）和页面。

1.2.2　搭建 Eclipse 开发环境

本书使用的Eclipse版本是Eclipse 2021-03，安装包可以从官方下载。

（1）创建新Maven项目。启动Eclipse，然后选择File→New→Maven Project，如图1.8所示。

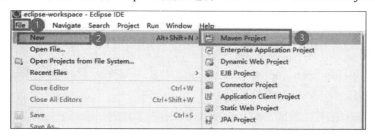

图 1.8　Eclipse 创建 Maven 项目

5

（2）使用默认配置，单击Next按钮，如图1.9所示。

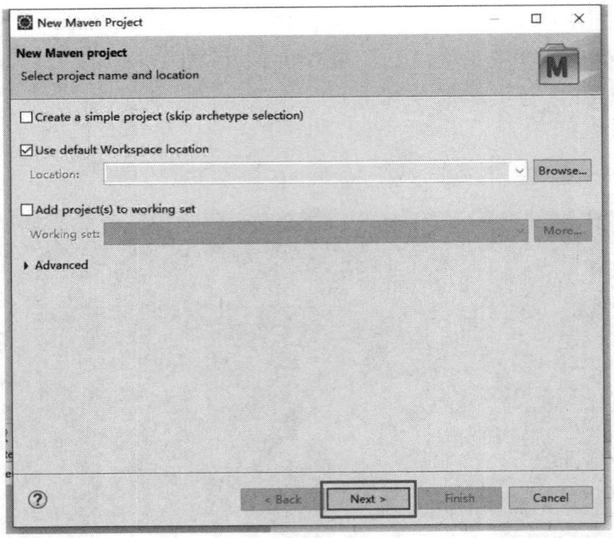

图1.9　Eclipse选择项目路径

（3）选择项目模板。Eclipse每次都会从Maven仓库拉取Archetype信息，默认是Maven中心仓库，网络如果比较慢，可以单击Configure...，如图1.10所示。

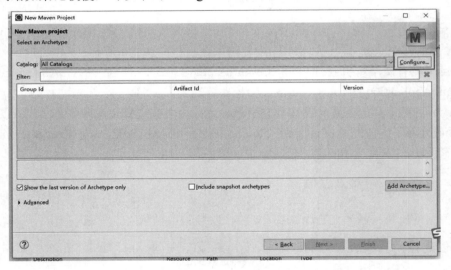

图1.10　Eclipse选择项目路径

这里我们配置华为云提供的Maven仓库，如图1.11所示，Catalog File填写https://mirrors.huaweicloud.com/repository/maven/，Description为仓库的名字，可以随意填写，这里填写为huawei-maven。单击OK按钮保存。

在Catalog中选择刚配置的huawei-maven，列表中加载出了Archetype信息，如图1.12所示，在Filter中输入org.springframework.boot，选择Artifact Id为spring-boot-sample-simple-archetype的条目，再单击Next按钮。

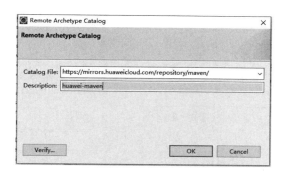

图 1.11　配置 Archetype Maven 仓库

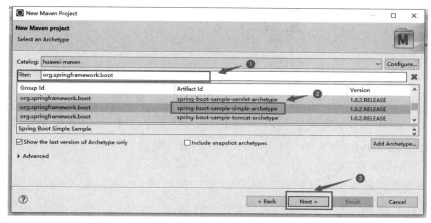

图 1.12　选择 spring-boot-sample-simple-archetype

（4）输入项目元信息。如图1.13所示，输入Group Id为com.example，Artifact Id为eclipsedemo，然后单击Finish按钮。

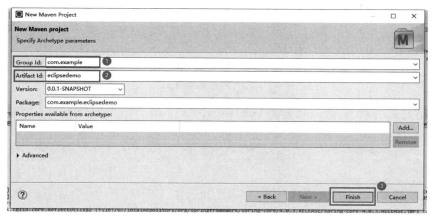

图 1.13　配置项目元信息

（5）此时已经完成搭建环境的所有操作。下面比较一下Eclipse和Spring Initializr生成的项目目录有哪些异同之处。

为了方便查看，设置包展示方式为Hierarchical，如图1.14所示。

在Project Explorer中展开所有目录，如图1.15所示。

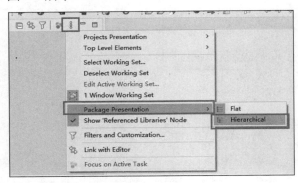

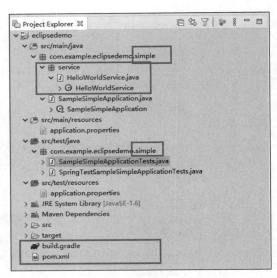

图1.14　修改Package展示方式

图1.15　Eclipse项目目录结构

通过对比可以看出，Eclipse与Spring Initializr生成的目录稍有区别，如生成的包都带有simple，代码生成了service包和HelloWorldService.java，构建工具除了生成pom.xml外，还生成了Gradle使用的配置文件build.gradle。

（6）接下来运行程序来验证搭建的开发环境可用。双击打开SampleSimpleApplication.java，选中main方法，右击弹出的快捷菜单，选择Run As→Java Application，如图1.16所示。

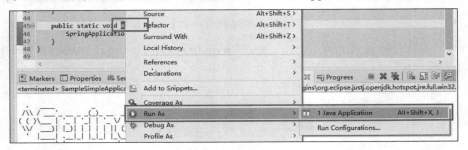

图1.16　运行程序主类

检查控制台的打印信息，查看到输出信息"Started SampleSimpleApplication in 1.629 seconds (JVM running for 2.234)"便说明搭建开发环境成功，如图1.17所示。

图1.17　项目运行的打印信息

1.2.3 搭建 IntelliJ IDEA 开发环境

本书使用的IntelliJ IDEA版本是社区版2020.3，下载地址为https://www.jetbrains.com/idea/download/#section=windows。

本节区别于1.2.2节中Eclipse使用Maven Archetype的构建方式，将IDEA和Spring Initializr相结合，使用IDEA加载1.2.1节中生成的初始项目。由于IDEA商业版中对Spring Initializr做了集成，因此商业版用户可以直接在IDEA中使用Spring Initializr创建项目，本书中不做演示。

（1）将项目文件夹放入IDEA工作目录。为了方便管理，将解压得到的demo目录放入IDEA的workspace目录。

注意　demo目录只有一个，也只有一级。

（2）使用IDEA加载demo项目。如图1.18所示，在左侧导航栏选择Projects，然后单击Open按钮。此时在IDEA右下角显示正在解析Maven依赖，如图1.19所示。Maven构建项目需要一些时间，等待构建完成。

图1.18　IDEA打开一个项目　　　　图1.19　项目运行的打印信息

（3）查看目录结构。在Project面板展开demo目录，如图1.20所示，相比于1.2.1节中的项目，会发现多了目录.idea和文件demo.iml，它们是IDEA用于管理项目生成的，我们在这里无须关心。

（4）查看项目的配置信息。双击pom.xml文件，在编辑窗口看到的文件内容如图1.21所示，可以看到我们在Spring Initializr中配置的信息。其中在parent标签内定义Spring Boot为父依赖，然后是项目元信息，之后在properties标签内定义了Java的版本。

然后再往下看，如图1.22所示，分别是spring web的starter、mybatis的starter和mysql驱动，这3个是我们在dependencies中选择的。后面的test starter是Spring Boot默认添加的，是测试所必需的依赖。

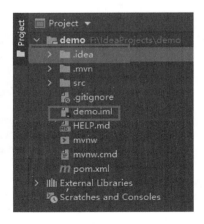

图1.20　IDEA中的项目目录

```xml
<parent>
    <groupId>org.springframework.boot</groupId>
    <artifactId>spring-boot-starter-parent</artifactId>
    <version>2.4.4</version>
    <relativePath/> <!-- lookup parent from repository -->
</parent>
<groupId>com.example</groupId>
<artifactId>demo</artifactId>
<version>0.0.1-SNAPSHOT</version>
<name>demo</name>
<description>Demo project for Spring Boot</description>
<properties>
    <java.version>11</java.version>
</properties>
```

图 1.21　POM.XML 文件内容 1

```xml
<dependencies>
    <dependency>
        <groupId>org.springframework.boot</groupId>
        <artifactId>spring-boot-starter-web</artifactId>
    </dependency>
    <dependency>
        <groupId>org.mybatis.spring.boot</groupId>
        <artifactId>mybatis-spring-boot-starter</artifactId>
        <version>2.1.4</version>
    </dependency>

    <dependency>
        <groupId>mysql</groupId>
        <artifactId>mysql-connector-java</artifactId>
        <scope>runtime</scope>
    </dependency>
    <dependency>
        <groupId>org.springframework.boot</groupId>
        <artifactId>spring-boot-starter-test</artifactId>
        <scope>test</scope>
    </dependency>
</dependencies>
```

图 1.22　POM.XML 文件内容 2

在文件最后，build标签内配置了maven插件，用来构建项目，如图1.23所示。

```xml
<build>
    <plugins>
        <plugin>
            <groupId>org.springframework.boot</groupId>
            <artifactId>spring-boot-maven-plugin</artifactId>
        </plugin>
    </plugins>
</build>
```

图 1.23　POM.XML 文件内容 3

以上这些都是Spring Initializr生成的，省去了无意义的复制、粘贴工作。所以在创建Spring Boot项目时，使用Spring Initializr是不错的选择。

（5）准备运行项目。由于MyBatis在项目启动时会连接数据库，我们先注释掉mybatis starter依赖，如图1.24所示。

图 1.24 注释掉 mybatis starter 依赖

选中要注释的行，使用快捷键Ctrl+/进行注释。

打开Maven面板，单击刷新按钮，使pom.xml的修改生效，如图1.25所示。

（6）找到DemoApplication.java文件，如图1.26所示，右击快捷菜单，选择Run。

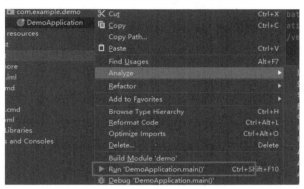

图 1.25　刷新 Maven 配置　　　　图 1.26　运行 DemoApplication.java

此时查看控制台，如图1.27所示，输出"Started DemoApplication in 6.48 seconds (JVM running for 7.487)"表示项目运行成功。

图 1.27　控制台输出信息

1.3　Spring Boot的依赖管理和自动配置

引入一个框架，通常需要引入依赖的JAR包和配置框架参数。单次引入框架并不烦琐，但是对于确定场景，依赖的JAR包和配置项往往是确定的，此时依赖管理和配置参数就成了重复性可程序化的操作。Spring Boot对于此等场景提供了Starter（启动器），简化了依赖管理和参数配置工作。本节将介绍starter相关的依赖管理和自动配置。

1.3.1 依赖管理

Spring Boot简化了Spring项目的依赖管理，我们需要了解这种简化是如何做到的。

（1）如果需要做Web开发，则只需要在项目pom.xml中配置web starter，代码如下：

```xml
<dependency>
  <groupId>org.springframework.boot</groupId>
  <artifactId>spring-boot-starter-web</artifactId>
</dependency>
```

这样项目便导入了所有Web开发需要的依赖。

（2）简单讲一下Starter的命名规则。Spring Boot官方要求starter都以spring-boot-starter开头，以场景（或模块）名结束，如上面的spring-boot-starter-web。如果是第三方提供的Starter，则应以第三方名称开头，以-spring-boot-starter结尾，例如mybatis-spring-boot-starter。

（3）Spring Boot在spring-boot-starter-parent的父项目spring-boot-dependencies中制定了默认版本号，代码如下：

```xml
<dependency>
  <groupId>org.springframework.boot</groupId>
  <artifactId>spring-boot-starter-web</artifactId>
  <version>2.4.4</version>
</dependency>
```

这里的dependency在dependencyManagement中定义，因此只有在项目中添加依赖时才生效，并且默认使用父项目中定义的版本号。

在spring-boot-starter-web的pom.xml文件中，可以看到其定义了如下依赖：

```xml
<dependencies>
   <dependency>
      <groupId>org.springframework.boot</groupId>
      <artifactId>spring-boot-starter</artifactId>
      <version>2.4.4</version>
      <scope>compile</scope>
   </dependency>
   <dependency>
      <groupId>org.springframework.boot</groupId>
      <artifactId>spring-boot-starter-json</artifactId>
      <version>2.4.4</version>
      <scope>compile</scope>
   </dependency>
   <dependency>
      <groupId>org.springframework.boot</groupId>
      <artifactId>spring-boot-starter-tomcat</artifactId>
      <version>2.4.4</version>
      <scope>compile</scope>
   </dependency>
   <dependency>
```

```xml
            <groupId>org.springframework</groupId>
            <artifactId>spring-web</artifactId>
            <version>5.3.5</version>
            <scope>compile</scope>
        </dependency>
        <dependency>
            <groupId>org.springframework</groupId>
            <artifactId>spring-webmvc</artifactId>
            <version>5.3.5</version>
            <scope>compile</scope>
        </dependency>
    </dependencies>
```

在这5个依赖中,有3个直接与Web开发相关,分别是spring-boot-starter-tomcat、spring-web和spring-webmvc。

- spring-boot-starter-tomcat 是引入内嵌的 Tomcat,从而不用打成 WAR 包也可以运行。从名字即可看出,它是 Spring Boot 官方提供的 Starter。
- spring-web 和 spring-webmvc 是 Spring 提供的 Web 模块。

可以看出,通过引入spring-boot-starter-web依赖,便引入了项目所用到的全部关于Web开发的JAR,这便是Spring Boot依赖管理的基本原理。

2.1节将会详细说明Spring Boot官方提供的所有Starter。

1.3.2 自动配置

引入框架后,常需要一些配置,如Spring注解需要配置包扫描路径、静态资源的存储路径、Web容器端口号、字符编码或日志输出目录等。Spring Boot在配置上的优化,一是为框架提供了通用的配置,在无个性化配置时甚至可以零配置启动项目;二是为项目配置提供了统一的入口,可以通过一个配置文件搞定所有配置。

在1.3.1节中能看到,Web模块引入了Spring框架,使用Spring框架注解时就要配置包扫描。Spring Boot默认将程序的启动类(@SpringBootApplication所修饰的类)所在的包作为扫描路径,因此开发时只要将代码放到在启动类所在的包下,就无须做扫描包的配置。因为包扫描路径无法静态统一指定,因此它是一个特例。如需配置扫描路径,可以通过在启动类上添加注解@ComponentScan来指定。

对于一般的参数,Spring Boot官网文档进行了说明,包括全部参数与默认值,网址为https://docs.spring.io/spring-boot/docs/2.4.4/reference/html/appendix-application-properties.html#common-application-properties。我们可以通过该网址查看配置的默认值。

比如有静态资源匹配模式,如图1.28所示。可以看到默认配置为"/**",所有请求都会通过静态资源处理。

图 1.28　配置参数静态资源匹配模式

再比如静态资源路径，如图1.29所示。默认路径有4个，分别是类路径下的/META-INF/resources/、/resources/、/static/和/public/。所以只要将静态资源放置到这些路径下，便无须另外配置。

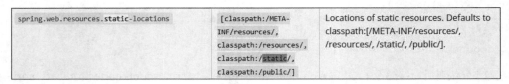

图 1.29　配置参数静态资源路径

以上是Spring Boot所做的自动配置，但在项目中仍会面临个性化的配置需求。设置项目参数，只需要在开发环境的resources目录下放置application.properties和application.yaml（application.yml）即可，Spring Boot可以自动识别这两个配置文件。Properties和YAML是Spring Boot支持的两种配置文件格式，YAML又可以写作YML。两种格式的语法都很简单，YAML的结构更清晰，占用空间也更优，因此本书中使用YAML。

【示例1.1】　修改Web服务端口号

在官方文档中找到端口的配置，如图1.30所示，默认是8080。

图 1.30　配置参数服务端口

将端口设置为9999，需要在application.yaml中添加代码：

```
server:
  port: 9999
```

再次启动程序会看到控制台输出"Tomcat started on port(s): 9999 (http) with context path"，表示端口已配置为9999。

【示例1.2】　修改Spring的banner

Spring启动时会在控制台打印banner，如图1.31所示。

修改banner的办法有两个：

- 一是将名称为 banner（JPG 或者 PNG）的图片放在 classpath 路径下。
- 二是修改 banner 路径。

这里我们采取第2种方式，将项目打印图标修改为如图1.32所示的样式，这个是Spring的图标。

将其保存为custom-banner.png，并放到resource/picture目录下。查看官方文档中banner的配置，如图1.33所示。

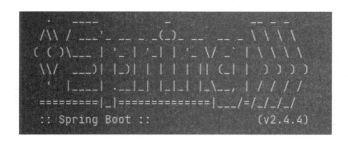

图 1.31　Spring 默认 banner

图 1.32　Spring 框架图标

| spring.banner.image.location | classpath:banner.gif | Banner image file location (jpg or png can also be used). |

图 1.33　配置参数 banner 图片路径

于是在application.yaml中添加如下配置：

```
spring:
  banner:
    image:
      location: /picture/custom-banner.png
```

然后重新启动项目，控制台输出如图1.34所示，表示配置成功。

图 1.34　控制台输出 Spring 图标

1.4　实战——Spring Boot版本的Hello World

使用Spring Boot构建一个Web项目，实现在浏览器中显示"Hello World"。

（1）使用Spring Initializr创建项目helloworld，并添加Spring Web依赖，如图1.35所示。保存项目到本地，并解压到工作空间目录。

（2）打开IDEA，选择File→Open菜单，如图1.36所示，选择helloworld文件夹，单击OK按钮。

（3）创建与HelloworldApplication.java同等级的controller包，在controller包中创建HelloWorldController.java文件，如图1.37所示。

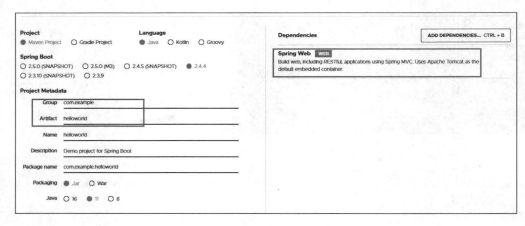

图 1.35 创建 helloworld 项目

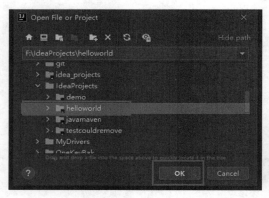

图 1.36 IDEA 打开 helloworld 项目

图 1.37 创建 HelloworldApplication 类

(4) 添加helloworld方法, 处理/helloworld请求。HelloWorldController.java代码如下:

```java
package com.example.helloworld.controller;

import org.springframework.stereotype.Controller;
import org.springframework.web.bind.annotation.RequestMapping;
import org.springframework.web.bind.annotation.ResponseBody;

@Controller
@ResponseBody
public class HelloWorldController{
    @RequestMapping("helloworld")
    public String helloworld(){
        return "Hello World!";
    }
}
```

(5) 运行HelloworldApplication.java的main方法, 程序启动, 默认端口为8080, 控制台输出如图1.38所示。

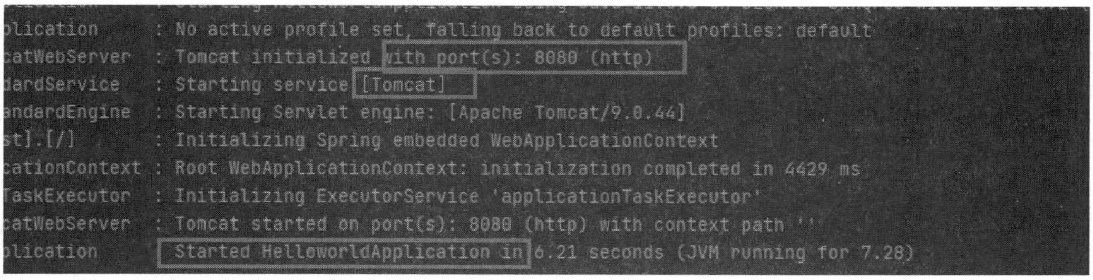

图 1.38　控制台输出信息

（6）打开浏览器，访问地址http://127.0.0.1:8080/helloworld，在页面显示出Hello World!，如图1.39所示。

图 1.39　浏览器输出 Hello World！

第 2 章

工程项目使用Spring Boot的步骤

本章将从项目开发的角度介绍Spring Boot，包括项目构建、程序目录、配置参数、程序运行以及开发者工具等部分。本章更多关注具体的某个功能，不会深入介绍细节，是比较粗粒度的介绍，后面的章节会越来越细致。

本章主要涉及的知识点有：

- 如何构建String Boot项目。
- Spring Boot默认的代码目录。
- Spring Boot提供的配置功能。
- @SpringBootApplication注解如何使用。
- 如何运行Spring Boot程序。
- Spring Boot开发者工具有哪些功能。

2.1 构建项目

本节将介绍构建工具的选择、依赖管理，以及Spring官方提供的Starter。希望通过本节的学习，读者能够自己选择项目构建工具，以及对Spring Boot提供的功能有更多的了解。

2.1.1 构建工具

构建工具有许多，常提到的有Ant、Maven和Gradle，Spring Boot官方建议使用Maven或Gradle，并且Spring Boot发布包会提交到Maven Central仓库。如果不适用Maven或Gradle，至少应该选择一款具有依赖管理功能的构建工具。

对于Maven和Gradle的使用，在本书中不会展开，本书所有示例均以Maven为例给出。

对于Maven和Gradle，Spring Boot官方均给出了其插件的使用手册，其网址在表2.1中给出。

表 2.1　构建工具插件文档地址

名　　称	网　　址
Spring Boot的Maven插件	https://docs.spring.io/spring-boot/docs/2.4.4/maven-plugin/reference/htmlsingle/
Spring Boot的Gradle插件	https://docs.spring.io/spring-boot/docs/2.4.4/gradle-plugin/reference/htmlsingle/

2.1.2　Starter、JAR 与依赖

为了方便添加依赖，Spring Boot提供了一个概念，称作Starter。Starter是对一个开发场景用到的所有依赖包的集中描述。当我们开发需要某一特定场景时，只需引入对应的Starter即可。

Starter的本质就是JAR包，对Starter的引用与其他JAR包的引用没有不同。在第1章中，我们曾用到过Starter引入Spring Web依赖，它的artifactId是spring-boot-starter-web，其实就是Spring Boot官方提供的一个Starter，对应的引用就是在项目pom.xml中添加dependency，代码如下：

```xml
<dependency>
    <groupId>org.springframework.boot</groupId>
    <artifactId>spring-boot-starter-web</artifactId>
</dependency>
```

在IDEA中，可以通过Ctrl+鼠标左键单击spring-boot-starter-web来打开这个Starter的POM文件，其文件名为spring-boot-starter-web-2.4.4.pom。在这个POM文件中，可以看到其中最重要的dependencies部分，代码如下：

```xml
<dependencies>
  <dependency>
    <groupId>org.springframework.boot</groupId>
    <artifactId>spring-boot-starter</artifactId>
    <version>2.4.4</version>
    <scope>compile</scope>
  </dependency>
  <dependency>
    <groupId>org.springframework.boot</groupId>
    <artifactId>spring-boot-starter-json</artifactId>
    <version>2.4.4</version>
    <scope>compile</scope>
  </dependency>
  <dependency>
    <groupId>org.springframework.boot</groupId>
    <artifactId>spring-boot-starter-tomcat</artifactId>
    <version>2.4.4</version>
    <scope>compile</scope>
  </dependency>
  <dependency>
    <groupId>org.springframework</groupId>
    <artifactId>spring-web</artifactId>
    <version>5.3.5</version>
    <scope>compile</scope>
  </dependency>
```

```xml
<dependency>
    <groupId>org.springframework</groupId>
    <artifactId>spring-webmvc</artifactId>
    <version>5.3.5</version>
    <scope>compile</scope>
</dependency>
</dependencies>
```

通过Maven的依赖传递机制，在dependencies中定义的这5个dependency都将在项目中生效。

关于Starter的命名，在第1章中做过详细说明，这里再复习一下。Spring Boot官方Starter以spring-boot-starter开头，第三方Starter以项目名称开头、以-spring-boot-starter结尾。

表2.2~表2.4是官方提供的Starter，了解官方提供的场景有助于我们在开发时做选择。

表2.2中介绍了对技术、依赖封装的Starter。

表 2.2 应用类 Starter

名 称	描 述
spring-boot-starter	核心Starter，包含自动配置、日志和YAML格式的配置文件的支持
spring-boot-starter-activemq	使用Apache ActiveMQ支持JMS通信的场景
spring-boot-starter-amqp	使用Spring AMQP和Rabbit MQ的场景
spring-boot-starter-aop	使用Spring AOP和AspectJ的面向切面编程场景
spring-boot-starter-artemis	使用Apache Artemis支持JMS通信的场景
spring-boot-starter-batch	使用Spring Batch的场景
spring-boot-starter-cache	使用Spring框架caching的场景
spring-boot-starter-data-cassandra	使用分布式数据库Cassandra和Spring Data Cassandra的场景
spring-boot-starter-data-cassandra-reactive	使用分布式数据库Cassandra和Spring Data Cassandra Reactive的场景
spring-boot-starter-data-couchbase	使用文档数据库Couchbase和Spring Data Couchbase的场景
spring-boot-starter-data-couchbase-reactive	使用文档数据库Couchbase和Spring Data Couchbase Reactive的场景
spring-boot-starter-data-elasticsearch	使用Elasticsearch搜索、分析引擎和Spring Data Elasticsearch的场景
spring-boot-starter-data-jdbc	使用Spring Data JDBC的场景
spring-boot-starter-data-jpa	使用Hibernate和Spring Data JPA整合的场景
spring-boot-starter-data-ldap	使用Spring Data LDAP的场景
spring-boot-starter-data-mongodb	使用文档数据库MongoDB和Spring Data MongoDB的场景
spring-boot-starter-data-mongodb-reactive	使用文档数据库MongoDB和Spring Data MongoDB Reactive的场景
spring-boot-starter-data-neo4j	使用Neo4j图数据库和Spring Data Neo4j的场景
spring-boot-starter-data-r2dbc	使用Spring Data R2DBC的场景
spring-boot-starter-data-redis	通过Spring Data Redis和Lettuce客户端来使用Redis键值对数据存储的场景
spring-boot-starter-data-redis-reactive	通过Spring Data Redis响应式和Lettuce客户端来使用Redis键值对数据存储的场景
spring-boot-starter-data-rest	通过Spring Data REST以REST的方式使用Spring Data存储的场景

（续表）

名称	描述
spring-boot-starter-data-solr	使用带有Spring Data Solr的Apache Solr搜索平台的场景，这个Starter从Spring Boot 2.3.9版本开始被标记为Deprecated
spring-boot-starter-freemarker	使用FreeMarker视图构建MVC Web应用的场景
spring-boot-starter-groovy-templates	使用Groovy Templates视图构建MVC Web应用的场景
spring-boot-starter-hateoas	使用Spring MVC和Spring HATEOAS来构建超媒体的RESTful Web应用
spring-boot-starter-integration	使用Spring Integration的场景
spring-boot-starter-jdbc	使用JDBC的场景，并使用HikariCP作为连接池
spring-boot-starter-jersey	使用JAX-RS和Jersey来构建RESTful Web应用的场景，这个Starter可以替代spring-boot-starter-web
spring-boot-starter-jooq	使用jOOQ来访问SQL数据库，这个Starter可以替代spring-boot-starter-data-jpa或spring-boot-starter-jdbc
spring-boot-starter-json	开启读写JSON的支持
spring-boot-starter-jta-atomikos	使用Atomikos支持JTA事务的场景
spring-boot-starter-jta-bitronix	使用Bitronix实现JTA事务的场景。这个Starter从Spring Boot 2.3.0开始标记为Deprecated
spring-boot-starter-log4j2	使用Log4j2来实现日志功能，是spring-boot-starter-logging的替代者
spring-boot-starter-logging	使用Logback来实现日志功能，也是Spring Boot的默认日志工具
spring-boot-starter-mail	使用Java Mail和Spring框架邮件功能的场景
spring-boot-starter-mustache	使用Mustache视图构建Web应用的场景
spring-boot-starter-oauth2-client	使用Spring Security's OAuth2或OpenID Connect客户端的场景
spring-boot-starter-oauth2-resource-server	使用Spring Security's OAuth2服务器的场景
spring-boot-starter-quartz	使用Quartz任务调取器的场景
spring-boot-starter-rsocket	用于构建RSocket客户端和服务的场景
spring-boot-starter-security	使用Spring Security的场景
spring-boot-starter-test	用于测试，提供了多个测试库，包括JUnit Jupiter、Hamcrest和Mockito
spring-boot-starter-thymeleaf	使用Thymeleaf视图来构建MVC Web应用的场景
spring-boot-starter-validation	使用Hibernate Validator来实现Java Bean Validation的场景
spring-boot-starter-web	用于Web开发场景，包含RESTful和Spring MVC，并且默认使用了内置的Tomcat
spring-boot-starter-web-services	使用Spring Web Services的场景
spring-boot-starter-webflux	使用Spring框架的Reactive Web模块来构建WebFlux应用的场景
spring-boot-starter-websocket	使用Spring框架的WebSocket模块来构建WebSocket应用的场景

表2.3介绍提供生产环境工具的Starter。

表 2.3 生产环境工具类 Starter

名称	描述
spring-boot-starter-actuator	此Starter提供用于应用监控和管理相关功能，这些功能在生产环境中非常实用

表2.4介绍Web容器相关Starter。

表 2.4　容器类 Starter

名 称	描 述
spring-boot-starter-jetty	使用Jetty作为Servlet容器，是spring-boot-starter-tomcat的替代者之一
spring-boot-starter-reactor-netty	使用Reactor Netty作为内置的响应式HTTP服务器
spring-boot-starter-tomcat	使用Tomcat作为Servlet容器，也是使用spring-boot-starter-web时的默认选项
spring-boot-starter-undertow	使用Under作为Servlet容器，是spring-boot-starter-tomcat的替代者之一

在网络上可以找到社区提供的Starters。Spring Boot项目的spring-boot-starters模块下有名为README的文件，在这个文件中有对社区提供的Starter说明。如果有必要，我们可以自己编写Starter提交到GitHub上。README文档可以在源代码中找到。

2.1.3　再说依赖管理

在1.3.1节我们就提到过依赖管理，并且引入了spring-boot-starter-web依赖。可以注意到，引入Web Starter的时候只提供了groupId和artifactId，没有提供必要的Version，这是因为Spring Boot为我们管理Web Starter的版本信息。

依赖的版本信息是在artifactId为spring-boot-dependencies的这一项目中管理的，它是spring-boot-starter-parent的父项目，而spring-boot-starter-parent又定义为我们的项目的父项目。对于spring-boot-starter-parent和spring-boot-dependencies，若是在Maven中，它们的packaging定义为"pom"，就是只有POM文件，没有代码。因为不是JAR包，所有不会在IDEA的External Libraries中找到。

我们可以在本地仓库中找到spring-boot-dependencies的POM文件（或者在IDEA中通过Ctrl+鼠标左键的方式打开），能够看到其中的内容主要集中在3个标签内，分别是properties、dependencyManagement和pluginManagement，如图2.1所示。

图 2.1　spring-boot-dependencies 的 POM 文件

（1）在 properties 中定义了非 Spring Boot 的 JAR 包的版本号，作为常量在dependencyManagement和pluginManagement中使用。properties中定义的一部分如图2.2所示，其命名格式都是"技术名称.version"。

图 2.2　properties 标签下的内容

（2）在dependencyManagement中定义了Spring官方框架的依赖，以及使用这些框架所必需的相应技术的依赖。dependencyManagement与dependencies不同，只在前者中定义不会在项目中生效，只有在项目的dependencies中定义后才会生效。以aspect相关的依赖为例，如图2.3所示，在version标签中，以"${aspectj.version}"对properties中定义的版本号进行引用。将版本号放到properties中定义的好处是其多次使用只需在一个位置定义，因此，我们需要修改版本号时，只需在项目中定义名称为aspectj.version即可。

另外，Spring Boot提供的JAR包的版本号是在dependencyManagement中定义的，如图2.4所示。这些版本都与spring-boot-starter-parent的版本保持一致。

图 2.3　使用 aspectj.version 属性的依赖　　　　图 2.4　Spring Boot 的 JAR 包版本指定

（3）pluginManagement的功能与dependencyManagement类似，其中定义的插件不会直接生效，只有当在项目中定义插件后才会生效，pluginManagement只是起到了版本管理的功能。

2.2　组　织　代　码

本节将介绍项目包的结构的规范，以及程序主类应该放置的包路径。

2.2.1　不建议使用 default package

在Java类中不进行package声明时，使用的是default package。在Spring Boot项目中，尤其使用到@ComponentScan、@ConfigurationPropertiesScan、@EntityScan、@SpringBootApplication 4个注解时，为了避免出现各种奇怪的问题，最好不要使用default package。

@SpringBootApplication注解中用到了@ComponentScan。@ComponentScan、@ConfigurationPropertiesScan和@EntityScan注解默认扫描所修饰类所在包下的所有类，若修饰类的包为default package，则将扫描所有JAR的所有类。如此一来，项目很可能无法启动。

即使没有用到这4个注解，根据Java编码规范，也要定义项目的包。

2.2.2 放置应用的 main 类

程序启动类最简单，也是Spring Boot官方推荐的方式，使用@SpringBootApplication注解修饰，并且将其放置在程序的root package下。

在第1章的helloworld项目实战中，程序启动类是HelloworldApplication.java。这个类放在包com.example.helloworld下。如果将这个类放到default package下，项目启动就会报找不到类而导致BeanDefinitionStoreException的错误，从而启动失败，错误信息如图2.5所示。进一步分析日志，会发现日志中多次提到程序使用了default package，如图2.6所示。

```
    ... 27 common frames omitted
Caused by: java.lang.ClassNotFoundException Create breakpoint : org.springframework.dao.DataAccessException <2 internal calls>
    at java.base/java.lang.ClassLoader.loadClass(ClassLoader.java:521) ~[na:na]
    ... 29 common frames omitted
```

图 2.5 使用默认包会造成的异常

```
** WARNING ** : Your ApplicationContext is unlikely to start due to a @ComponentScan of the default package.


2021-04-01 07:28:37.180  WARN 9296 --- [           main] ConfigServletWebServerApplicationContext : Exception encountered during context initialization
 - cancelling refresh attempt: org.springframework.beans.factory.BeanDefinitionStoreException: Failed to read candidate component class: URL
[jar:file:G:\localRepository\org\springframework\boot\spring-boot-autoconfigure\2.4.4\spring-boot-autoconfigure-2.4.4
.jar!/org/springframework/boot/autoconfigure/jdbc/DataSourceAutoConfiguration$EmbeddedDatabaseConfiguration.class]; nested exception is java.lang
.IllegalStateException: Could not evaluate condition on org.springframework.boot.autoconfigure.jdbc
.DataSourceAutoConfiguration$EmbeddedDatabaseConfiguration due to org/springframework/dao/DataAccessException not found. Make sure your own
configuration does not rely on that class. This can also happen if you are @ComponentScanning a springframework package (e.g. if you put a
@ComponentScan in the default package by mistake)
```

图 2.6 异常信息指明 default package

但是，如果为@ComponentScan指定扫描包的路径，这里需要将@SpringBootApplication替换掉，如图2.7所示，程序又能正常运行。虽然可以修复错误，但是为了避免不必要的麻烦，应该遵循代码规范，不要使用default package。

```
@SpringBootConfiguration
@EnableAutoConfiguration
@ComponentScan("com.example.helloworld.controller")
public class HelloworldApplication {
```

图 2.7 @ComponentScan 指定扫描包的路径

2.3 配 置 类

虽然可以用XML文件来配置，但是Spring Boot更推荐使用基于Java代码的配置方式。最好使用单个被@Configuration修饰的类作为主要配置源，而在2.2节中提到的程序启动类就是这一角色的最佳选项。

2.3.1 导入其他配置类

前面提到要有单个主要配置类，但并不是说所有的配置都要放在一个类中。定义其他配置类只需要用@Configuration来修饰即可。然后需要使程序识别配置类，这时有两个注解可选：

- 一是在使用@Import 时将配置类导入。
- 二是使用@ComponentScan 扫描。

2.3.2　导入 XML 配置

使用XML文件实现配置虽然不推荐，但也是可行的。在使用XML配置时，可以在@Configuration修饰的类中使用@ImportResource注解，指定需要加载的XML文件。

2.4　再说自动配置

自动配置功能是指Spring Boot根据我们添加的JAR包依赖自动生效必要的配置。通过在@Configuration注解修饰的类上添加@EnableAutoConfiguration注解或者@SpringBootApplication注解，可以开启Spring Boot的自动配置功能。因为@SpringBootApplication注解包含@EnableAutoConfiguration注解，所以在使用时要二选一。通常在程序启动类上添加自动配置注解@SpringBootApplication，这也和我们在第1章中用Spring Initializr生成的项目写法一致。

2.4.1　用户配置替换自动配置

当我们提供自定义的配置时，Spring Boot的相关自动配置将不会生效。例如，当我们提供DataSource配置时，Spring Boot默认的内置数据库就不会再生效。

通过程序参数"--debug"，使程序启动时输出Spring Boot自动配置的所有信息。如图2.8所示，单击IDEA右上角的Select Run/Debug Configuration组件的下拉按钮，然后选择Edit Configurations，打开Select Run/Debug Configuration窗口。在如图2.9所示的选框位置录入"--debug"，单击OK按钮。

图 2.8　编辑启动配置

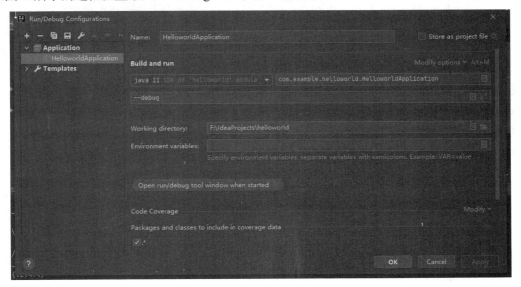

图 2.9　在配置中添加"--debug"参数

启动我们的项目，会看到控制台输出多了一部分 CONDITIONS EVALUATION REPORT，格式如图2.10 所示。

在CONDITIONS EVALUATION REPORT下分为 4部分：Positive matches、Negative matches、Exclusions 和Unconditional classes，分别对应着生效的自动配置、 不生效的自动配置、指定不生效的自动配置和无条件 生效的自动配置。

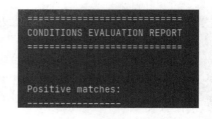

图 2.10 CONDITIONS EVALUATION REPORT 日志

对于debug这个参数设置，上面使用"--debug"是一种方式，另外还可以在配置文件中配置"debug=true"来实现。例如，在application.yaml中设置debug: true。

2.4.2 指定禁用生效的自动配置类

通过@SpringBootApplication注解和@EnableAutoConfiguration注解的两个参数exclude和 excludeName来指定禁止生效的自动配置类。例如，使用@SpringBootApplication注解的exclude 参数指定禁止MultipartAutoConfiguration这一自动配置，如图2.11所示。

图 2.11　指定禁用 MultipartAutoConfiguration 自动配置

这里已经配置了debug参数，启动项目，在控制台输出中，Exclusions显示有 MultipartAutoConfiguration这一项，如图2.12所示。

图 2.12　日志打印禁用信息

excludeName和exclude参数的功能是相同的，只是excludeName指定类的完全限定名，方便在被指定类不在classpath中时使用。

除了通过注解外，也可以通过在配置文件中配置参数spring.autoconfigure.exclude来指定不生效的自动配置类。

2.5　Spring Bean与依赖注入

在Spring Boot项目中，我们能够轻松地使用标准的Spring框架相关技术来创建Bean和注入其依赖。本节将结合示例讲解与创建Bean和依赖注入相关的Spring注解。

创建Spring Bean需要指定扫描路径，可以使用@ComponentScan注解或@SpringBootApplication

注解。如果按照2.2节中的组织代码规范，则框架会自动识别需要扫描的路径，无须指定路径参数。

在1.4节的helloworld项目实战中，我们用到了@SpringBootApplication注解与@Controller注解，并且由于规范的代码路径，因此在没有指定包扫描路径的前提下，由注解@Controller修饰的HelloWorldController类也被加载到容器中。

定义Bean的注解除了@Controller外，还有@Component、@Service和@Repository，这些都可以自动地注册组件到Spring容器中。

对于Bean的依赖注入，常用的注解是@Autowired。通过下面的示例来演示@Autowired注解的使用。

【示例2.1】 在1.4节helloworld实战项目的基础上添加Service层

（1）创建com.example.helloworld.service包，注意新建的service包与controller包在同一层级，如图2.13所示。

图 2.13　创建 service 包

（2）在service包下创建Java类，类名为HelloWorldService.java，并在类上添加注解@Service，如图2.14所示。

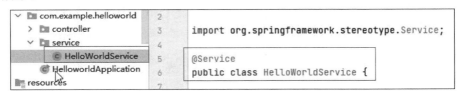

图 2.14　添加@Service 注解

在HelloWorldService.java类中添加方法helloWorldTimeStamp，获取当前时间戳并返回，代码如下：

```
public String helloWorldTimeStamp(){
    return "service:"+String.valueOf(System.currentTimeMillis());
}
```

（3）为HelloWorldController添加属性，引入对HelloWorldService的依赖，并使用@Autowired为字段标注，代码如下：

```
@Autowired
HelloWorldService helloWorldService;
```

再来修改helloworld方法，调用helloWorldService的方法helloWorldTimeStamp，将结果拼接到Hello World之后，增加的代码如图2.15所示。

```
@RequestMapping("helloworld")
public String helloworld(){
    return "Hello World!" + helloWorldService.helloWorldTimeStamp();
}
```

图 2.15　添加时间戳信息

（4）启动程序，待程序运行后，在浏览器访问http://127.0.0.1:8080/helloworld，效果如图2.16所示。注意，这里使用的是默认端口8080。

程序运行无误，说明添加Service层与自动注入都已正确实现。

图 2.16　在浏览器查看效果

@Autowired注解的使用方法有许多种，上面的示例中在字段上使用，还可以在构造器上使用，如图2.17所示。在只有一个构造器时，可以省略@Autowired注解。

```
HelloWorldService helloWorldService;
@Autowired
public HelloWorldController(HelloWorldService helloWorldService) {
    this.helloWorldService = helloWorldService;
}
```

图 2.17　在构造器上使用@Autowired 注解

 使用构造器注入时需要注意，一旦创建Bean后，注入关系便无法修改。

2.6　使用@SpringBootApplication注解

在前面的章节涉及自动配置和依赖管理时，曾多次提及@SpringBootApplication注解。本节将更详细地介绍@SpringBootApplication注解的使用。

我们第一次使用@SpringBootApplication注解是在第1章，在使用Spring Initializr生成的代码中。这个注解用来修饰程序启动类，当我们运行main方法时，便启动了一个Spring容器。@SpringBootApplication注解之所以功能强大，是因为它集成了其他注解，也被称为复合注解。

@SpringBootApplication注解集成的注解分别是@EnableAutoConfiguration、@ComponentScan和@Configuration。也就是说，在使用@SpringBootApplication注解时，相当于使用@EnableAutoConfiguration、@ComponentScan和@Configuration共同修饰了程序启动类。将程序启动类的@SpringBootApplication替换成这3个注解，程序同样可以正常运行。对于这3个注解的功能，在前面的章节中都有讲述，在此不再赘述。

在@SpringBootApplication注解中，为@EnableAutoConfiguration、@ComponentScan和@Configuration注解的属性定义了别名，因此在@SpringBootApplication注解上可以使用别名赋值。

2.7　运 行 程 序

Spring Boot支持内置Http Server，使得程序运行非常方便。本节将介绍在IDEA中以及打包后如何运行Spring Boot程序。

2.7.1 在 IDE 中运行

在IDE中运行程序，通常是在开发状态。在IDEA中运行程序，在第1章和本章的前几节中使用过，其方法就是选中程序启动类的main方法并右击，在弹出的菜单中选择运行的条目，如图2.18所示。

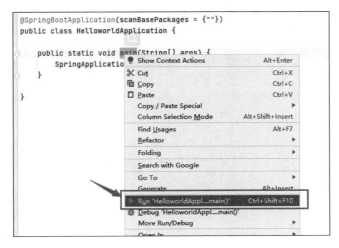

图 2.18　右击运行程序

在Eclipse中运行程序与IDEA操作非常类似，如图2.19所示。

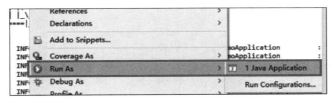

图 2.19　在 Eclipse 中右击运行

2.7.2 打成 JAR 包运行

首先，需要将项目打成可执行的JAR包，在这里使用Maven实现。

（1）在IDEA中，打开Maven面板，如图2.20所示，然后双击Plugins下的jar:jar进行打包。待打包完成，项目target目录下会生成JAR包。

（2）打开IDEA的Terminal面板，如图2.21所示。

（3）此时默认位置是项目路径下，然后运行JAR包，执行如下命令：

```
java -jar target/helloworld-0.0.1-SNAPSHOT.jar
```

如果要开启远程调试，需要在命令上添加参数"-Xdebug -Xrunjdwp:server=y, transport=dt_socket,address=8000,suspend=n"，命令如下：

```
java -Xdebug -Xrunjdwp:server=y,transport=dt_socket,address=8000,suspend=n -jar target/helloworld-0.0.1-SNAPSHOT.jar
```

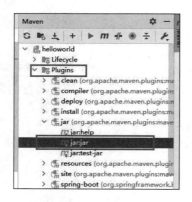

图 2.20　使用 Maven 插件打包

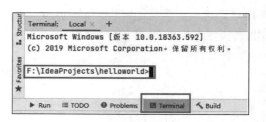

图 2.21　在 IDEA 中使用 Terminal

2.7.3　使用 Maven 插件运行

可以使用Spring Boot Maven插件的run goal快速编译并运行程序。这么做，运行的代码是target目录下零散的class文件，而不是JAR包，这个操作和在IDEA中在main方法上右击运行一样。

通过Spring Boot Maven插件可以使用IDEA中Maven面板的spring-boot:run，如图2.22所示。

也可以像2.7.2节中那样打开Terminal终端，在项目根目录下使用如下命令：

```
mvn spring-boot:run
```

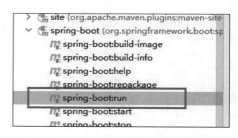

图 2.22　使用 Maven 的 Spring Boot 插件

2.7.4　使用 Gradle 插件运行

Spring Boot Gradle插件提供了相似的功能，通过bootRun Task来实现，只需要在Terminal中运行命令即可：

```
gradle bootRun
```

2.7.5　热部署

JVM热部署技术和使用JVM插件Jrebel，都可以支持Spring Boot项目热部署，但是使用起来相对复杂。spring-boot-devtools模块支持应用快速重新启动，虽然不是热部署，但简单易用，更多的内容在下一节中介绍。

2.8　开发者工具

在上一节的最后提到了spring-boot-devtools模块的快速重启功能，spring-boot-devtools模块还提供了一些开发时会用到的功能，这些功能可以方便我们快速开发。

首先，需要导入对spring-boot-devtools模块的依赖，使用Maven代码如下：

```
<dependencies>
  <dependency>
    <groupId>org.springframework.boot</groupId>
    <artifactId>spring-boot-devtools</artifactId>
    <optional>true</optional>
  </dependency>
</dependencies>
```

使用Gradle代码如下：

```
dependencies {
  developmentOnly("org.springframework.boot:spring-boot-devtools")
}
```

在Maven中设置optional为true，在Gradle中使用了developmentOnly，这些设置会使得在打包时默认不包括spring-boot-devtools模块，以及依赖此项目的项目不会继承spring-boot-devtools模块。为了在JAR包中包含spring-boot-devtools模块，需要在Maven中设置excludeDevtools属性为false，或在Gradle中设置打包时包含developmentOnly插件。

当以JAR包方式运行或者以特定类加载器加载时，开发者工具会被禁用。将参数spring.devtools.restart.enabled设置为true可以启动开发者工具。但要注意，开启开发者工具会有安全风险，所以可以在生产环境的配置文件中，显式设置属性spring.devtools.restart.enabled为false，确保禁用开发者工具。

2.8.1 默认配置

spring-boot-devtools模块为Spring的一些框架做了默认配置。比如，在使用Spring MVC或者Spring WebFlux开发时，开发者工具会自动开启DEBUG级别日志输出。再进一步，可以通过spring.mvc.log-request-details和spring.codec.log-request-details来获得更多的输出信息。

再比如，Spring的某些框架使用了缓存机制。缓存机制在生产环境中能有效提升性能，但是在开发环境中可能会导致我们的修改无法立即生效，从而带来不必要的麻烦。鉴于此，spring-boot-devtools模块默认关闭缓存机制。

如果不需要spring-boot-devtools模块的默认配置，可以通过设置参数spring.devtools.addproperties为false来关闭默认配置。

spring-boot-devtools模块的全部默认配置可以在源代码spring-boot-devtools/src/main/java/org/springframework/boot/devtools/env/DevToolsPropertyDefaultsPostProcessor.java中找到。

2.8.2 自动重启

使用spring-boot-devtools模块监听文件状态，在文件修改后项目自动重启，这个功能可以使得代码变化立即生效，开发调试时非常实用。

触发项目重启，在不同IDE中有不同的实现。在Eclipse中，修改文件内容后的保存操作会触发重启；在IDEA中，构建项目完成时触发重启；如果使用的是构建工具Maven或Gradle，那么执行命令"mvn compile"或"gradle build"将会触发重启。

对于使用构建工具Maven或Gradle的场景，禁用forking会导致自动重启功能无法执行。

1. 记录自动配置的变化

默认情况下，每次重启，在日志中都会输出condition evaluation有关打印程序自动配置的变动信息。通过将属性spring.devtools.restart.log-condition-evaluation-delta设置为false，可以禁用condition evaluation日志。

2. 指定需要触发重启的目录

默认情况下，开发者工具会监控classpath除静态目录（这些静态目录会在下面说明）之外的文件的变化，也支持我们添加classpath以外目录的监控。

使用属性spring.devtools.restart.additional-paths可以指定classpath以外的目录，当目录文件变化时，触发重启或重新加载。

3. 修改默认不触发重启目录配置

在静态资源以及一些视图模板等目录中，文件的变动通常不需要重启程序。默认情况下，/META-INF/maven、/META-INF/resources、/resources、/static、/public和/templates目录下文件的变动不会触发程序重启。

当这些默认目录需要被监控时，剔除默认目录配置，通过属性spring.devtools.restart.exclude来实现。例如，我们需要/static和/public目录触发重启，在配置文件application.yaml中配置如下：

```
spring:
  devtools:
    restart:
      exclude: "static/**,public/**"
```

上面的方式对默认配置做了修改，也可以使用属性spring.devtools.restart.additional-exclude剔除默认配置以外的目录，这种方式也适用于属性spring.devtools.restart.additional-paths指定的目录。

4. 禁用重启功能

在application.yaml中添加配置，设置属性spring.devtools.restart.enabled为false，可以禁用重启功能。注意，修改的配置在重启一次后才生效。

如果需要对所有库都完全禁用重启功能，则需要在SpringApplication.run(…)代码之前将spring.devtools.restart.enabled设置到系统属性中，代码如下：

```
public static void main(String[] args) {
    System.setProperty("spring.devtools.restart.enabled", "false");
    SpringApplication.run(MyApp.class, args);
}
```

5. 指定触发器文件

开发所使用的IDE很可能会在代码每次修改后实时编译，而我们只需要在某个特定时刻重启，这时可以指定一个文件作为"触发器"，只有当这个文件被修改时才触发重启。注意每次文件修改仍然会触发检查，但只有触发器文件被修改才会重启。

设置属性spring.devtools.restart.trigger-file，为其指定一个类路径下的文件，不可以是目录。比如在resources目录下名为.reloadtrigger的文件，在application.yaml中添加配置如下：

```
spring:
  devtools:
    restart:
      trigger-file=:
        reloadtrigger: .reloadtrigger
```

这里的相对路径是"项目根目录/target/classes"。

此时，只有当.reloadtrigger变化后，如果是在IDEA中，则在build后才会重启。

6. 自定义重启加载项

Spring Boot通过两个类加载器来实现快速重启，这两个类加载器分别称作base类加载器和restart类加载器。其中，base类加载器负责加载不发生变化的字节码文件，如在JAR包中的字节码文件；restart类加载器负责加载开发中的字节码文件。在触发重启后，restart类加载器实例以及其加载的内容会被丢弃，重新创建一个restart类加载器实例。

在默认情况下，IDEA打开的开发项目中所有代码会被restart类加载器加载。但是有时候，默认加载方式下重启速度不够快，比如多模块项目重新加载了不会变化的模块。

开发者工具支持我们自定义需要或者不需要被重新加载的类。首先，创建文件META-INF/spring-devtools.yaml，在这个文件中通过两个属性restart.include和restart.exclude指定是否被重新加载。include指由restart类加载器加载，exclude指由base类加载器加载。这两个属性的键必须唯一，值支持正则匹配，示例如下：

```
restart:
  exclude:
    companycommonlibs: "/mycorp-common-[\\w\\d-\\.]+\\.jar"
  include:
    projectcommon: "/mycorp-myproj-[\\w\\d-\\.]+\\.jar"
```

类路径下的所有META-INF/spring-devtools.yaml或META-INF/spring-devtools.properties都会生效，包括JAR包中的文件。

7. 局限性

自动重启功能不支持使用标准API ObjectInputStream进行反序列化的对象，我们可以使用Spring提供的ConfigurableObjectInputStream和Thread.currentThread().getContextClassLoader()来替代。

但是，对于不支持上下文类加载器的第三方库的反序列化，或许就只能联系其作者来解决了。

2.8.3　使用 LiveReload 自动刷新

spring-boot-devtools模块内置了LiveReload服务，这是一个可以在资源变动后触发浏览器刷新的应用。浏览器需要安装LiveReload插件。

使用属性spring.devtools.livereload.enabled来配置LiveReload服务的开启或关闭。

但是，要注意两点，一是启动多个LiveReload服务时，只有第一个生效；二是开启自动重启功能是开启自动刷新功能的前提。

2.8.4　全局设置

通过在目录"$HOME/"添加配置文件可以为在某一台计算机上运行的所有使用开发者工具的Spring Boot程序提供配置参数。配置目录为$HOME/.config/spring-boot，在此目录下添加名称为spring-boot-devtools.properties、spring-boot-devtools.yaml或spring-boot-devtools.yml的配置文件，配置的内容格式和在application.yaml中的相同。

老版本识别的目录文件是$HOME/.spring-bootdevtools.properties，新版本为兼容老版本，在配置目录$HOME/.config/spring-boot下找不到配置文件时，会去找$HOME/.spring-bootdevtools.properties。

在上面的配置文件中，不支持使用环境变量。

2.9　打包应用到生产环境

Spring Boot程序在生产环境部署时非常方便，只需打成JAR包即可。使用构建工具可以方便地打包，比如使用Maven，在项目根目录下执行命令"mvn package"，等待程序执行完，在target目录下生成程序的可执行JAR包。

此外，spring-boot-actuator模块提供了程序监控、请求监控统计等实用的功能，如果需要，可以引入此模块。

2.10　实战——使用Maven创建完整的工程项目

在了解了Spring Boot项目结构和默认配置之后，本节通过Maven构建项目，然后手动搭建目录结构。

（1）打开IDEA，单击File→New→Project菜单，如图2.23所示。

第 2 章　工程项目使用 Spring Boot 的步骤

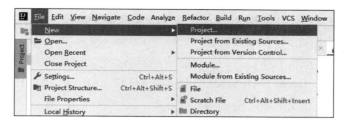

图 2.23　新建项目

（2）在New Project窗口选择Maven，然后单击Next按钮，如图2.24所示。

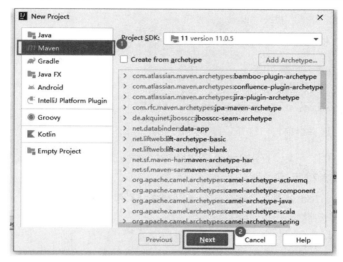

图 2.24　选择使用 Maven

（3）在Name中录入项目名称，这里命名为myproject。单击Artifact Coordinates查看项目元信息，这里无须修改，如图2.25所示。单击Finish按钮，完成项目信息录入。

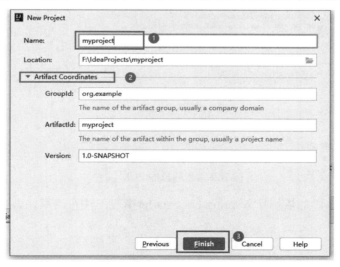

图 2.25　录入项目元信息

（4）最后在IDEA弹出的Open Project窗口中单击This Window，选择使用当前窗口打开新创建的Maven项目，如图2.26所示。

（5）等待IDEA打开当前项目之后，可以看到Maven已经为我们创建好基本的项目结构，如图2.27所示。其中，".idea"是IDEA工具创建的目录，myproject.iml文件用于存放IDEA与项目相关的配置文件，开发中通常用不到此目录。

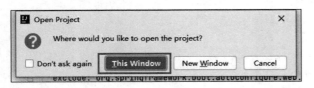
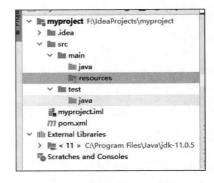

图 2.26　选择当前窗口打开项目　　　　图 2.27　新项目的目录结构

打开pom.xml文件，可以看到Maven在POM文件中已经写入项目基本的元信息，并制定了编译器的版本，如图2.28所示。

（6）添加Spring Boot依赖。首先，为当前项目指定父项目为spring-boot-starter-parent，在pom.xml中添加如下代码：

```xml
<parent>
    <groupId>org.springframework.boot</groupId>
    <artifactId>spring-boot-starter-parent</artifactId>
    <version>2.4.4</version>
</parent>
```

图 2.28　新项目的 POM 文件

然后，需要通过添加依赖导入spring-boot-starter模块，注意这里包含<dependencies>标签，代码如下：

```xml
<dependencies>
    <dependency>
        <groupId>org.springframework.boot</groupId>
```

```
            <artifactId>spring-boot-starter</artifactId>
        </dependency>
</dependencies>
```

最后，使用Maven面板上的刷新按钮让Maven配置生效，如图2.29所示。

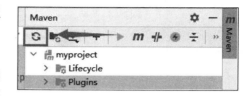

（7）添加程序启动类。根据在2.2节中的讲解，需要先创建项目的基础包，这里我们创建com.example.myproject包。在目录main/java上右击，在弹出的菜单中选择New→Package，如图2.30所示。

图 2.29　刷新 Maven

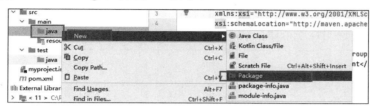

图 2.30　新建包

录入com.example.myproject，如图2.31所示，按回车键确认。

在刚创建的包com.example.myproject上右击，在弹出的菜单中选择New→Java Class，如图2.32所示。

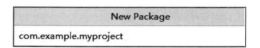

图 2.31　录入包名

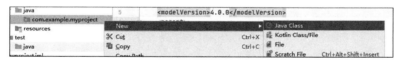

图 2.32　创建 Java 类

录入类名为MyApp，如图2.33所示，按回车键确认。

在生成的MyApp.java文件中，给MyApp Class添加Spring Boot注解@SpringBootApplication，然后将光标置于@SpringBootApplication后面，按快捷键Alt+Enter，在提示菜单中选择Import class，如图2.34所示。

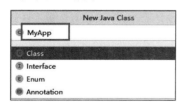

图 2.33　录入类名

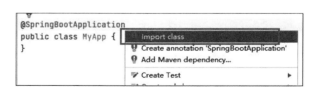

图 2.34　导入包名

然后，添加main方法，代码如下：

```
public static void main(String[] args) {
    SpringApplication.run(MyApp.class, args);
}
```

同样，需要为SpringApplication导入包名。至此，完成Spring Boot项目的搭建工作。

（8）启动项目，检验成果。单击main方法，然后单击窗口左侧的绿色三角形，弹出的菜单中选择Run 'MyApp.main()'，如图2.35所示。

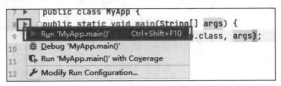

图2.35　启动程序

在控制台看到Spring Logo图片，以及Started MyApp in 2.303 seconds等信息，并无报错信息，如图2.36所示，说明程序搭建成功。

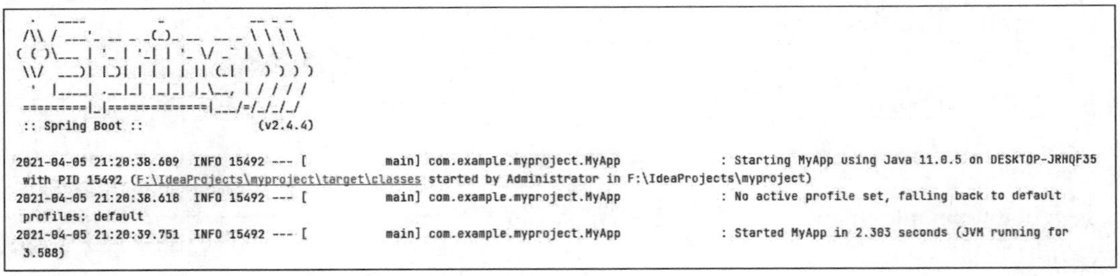

图2.36　程序成功运行的日志

在此次实战中，为了突出项目骨架，甚至没有导入对test的依赖。读者可以对比我们在这里搭建的项目与使用Spring Initializr创建的项目的区别，从而更透彻地理解Spring Boot的使用。

第 3 章

使用Spring Boot进行Web开发

Web开发是Java应用的重要方向之一，Spring Boot对于Web开发有着良好的支持，使用Spring Boot能够方便地整合Spring框架的MVC等Web模块，并且能够方便地集成Web相关框架。本章将介绍Spring Boot与Web开发相关的一些应用。

本章主要涉及的知识点有：

- Spring Boot对模板引擎的支持。
- 在Spring Boot中使用Thymeleaf开发页面。
- 在Spring Boot中使用Spring框架的文件上传功能。
- 整合Quartz Scheduler框架开发定时任务。
- 在Spring Boot中启用发送邮件功能。
- 在Spring Boot中使用Shiro进行认证和授权。

3.1 模板引擎

在Spring Boot开发的Web项目中，也可以使用Spring MVC框架支持的模板引擎技术进行动态HTML开发。动态HTML可以让网页界面与数据分离，业务代码与逻辑代码分离，从而提升开发效率，良好的设计使得代码重用变得更加容易。

Spring MVC支持多种模板引擎技术，如JSP、Thymeleaf、Freemarker、Velocity、Groovy、和Mustache等。Spring Boot官方提供了这4种技术的Starter：Thymeleaf、Freemarker、Groovy、和Mustache。在下一节中将通过示例简单介绍Thymeleaf的使用。

为什么不使用JSP？Spring Boot不推荐JSP，原因有两个：

- 并非所有Servlet容器都支持JSP，如容器Undertow不支持JSP。
- 对打包方式有限制，JAR包不支持JSP，若使用JSP，则需要打成WAR包。

3.2 使用Thymeleaf开发示例

Spring Boot为Thymeleaf提供了Starter：spring-boot-starter-thymeleaf。因此，需要添加依赖以开启对Thymeleaf的支持。通过下面4步搭建使用Thymeleaf开发的框架：

（1）在使用Thymeleaf时，首先需要在pom文件中引入该Starter，代码如下：

```
<dependency>
  <groupId>org.springframework.boot</groupId>
  <artifactId>spring-boot-starter-thymeleaf</artifactId>
</dependency>
```

修改POM配置后再刷新POM文件来使变更生效。

默认在/classpath:/templates目录下，后缀为.html的文件被识别为模板文件，由模板引擎来解析。目录/classpath:/templates对应的开发目录是src/main/resources/templates。

因此，在目录src/main/resources/templates下创建HTML文件作为页面。

（2）在目录src/main/resources/templates下创建名为thymeleaf.html的文件，并在文件中录入以下代码：

```
<!DOCTYPE html>
<html lang="en">
<head>
    <meta charset="UTF-8">
    <title>Thymeleaf</title>
</head>
<body>
    <h1>Using Thymeleaf</h1>
</body>
</html>
```

（3）为查看页面，在Controller中添加方法，处理请求"/thymeleaf"，代码如下：

```
@RequestMapping("thymeleaf")
public String thymeleaf(){
    return "thymeleaf";
}
```

Controller类上不能使用注解@RestController，应该使用注解@Controller，以及方法上不能使用注解@ResponseBody，这些是易错点。

（4）启动项目，在浏览器中访问http://127.0.0.1:8080/thymeleaf，效果如图3.1所示，这就是文件thymeleaf.html的展示效果。这里的端口要和在application.yaml中配置的端口（server.port）一致。

图 3.1 页面显示 Using Thymeleaf

此时,虽然在文件thymeleaf.html中尚未使用Thymeleaf的语法,但是开发动态HTML的流程已经走通。接下来,通过修改HTML代码在其中添加Thymeleaf的语法特性,然后在浏览器中查看效果来体验Thymeleaf的使用。

1. 获取 Java 对象的基本类型和字符串

(1)回到添加thymeleaf方法的Controller类中,在thymeleaf方法中添加变量。首先,为方法添加一个Map类型的参数,这个参数会被放到请求域中。然后,为map添加一个字符串类型变量,不妨命名为studentName,以及添加一个整型变量age,这两个变量将在HTML文件中获取,代码如下:

```
@RequestMapping("thymeleaf")
public String thymeleaf(Map<String, Object> map){
    map.put("studentName","张三");
    map.put("age", 16);
    return "thymeleaf";
}
```

(2)修改thymeleaf.html文件,在已有的代码<h1>标记下先添加<hr>,用来分隔代码,再添加一个<div>标签,并添加一个属性"th:text="${studentName}"",完整代码如下:

```
<!DOCTYPE html>
<html lang="en">
<head>
    <meta charset="UTF-8">
    <title>Thymeleaf</title>
</head>
<body>
    <h1>Using Thymeleaf</h1>
    <hr>
    <div th:text="${studentName}"></div>
</body>
</html>
```

(3)重启项目,或者按照第2章中Spring Boot开发者工具的说明,只需构建项目。打开浏览器,访问http://127.0.0.1:8080/thymeleaf,效果如图3.2所示。

图 3.2 页面更新后的效果

(4)分析涉及的语法。其中,th为Thymeleaf的命名空间;${}用于获取其中变量的值,这里就是获取studentName的变量值,其值为"张三";th:text表示将div中的内容替换成获取到的值。因此,<div th:text="${studentName}"></div>在经过Thymeleaf解析后,程序返回给浏览器的字符为<div>张三</div>。在浏览器页面中右击,选择"查看网页源代码",如图3.3所示。在新标签页中可以查看到源代码,如图3.4所示。

(5)结合上面分析的语法,追加一个<div>,以显示age的信息,代码如下:

```
<div th:text="${age}"></div>
```

最后,构建项目,浏览器中刷新页面,效果如图3.5所示,页面已经展示出age的信息。

图 3.3　查看网页源代码　　　　　　　　图 3.4　网页源代码

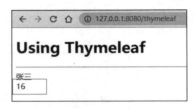

图 3.5　显示 age

2. 修改属性值

在第一个<div>中添加属性值th:name="${studentName}"，代码如下：

```
<div th:text="${studentName}" th:name="${studentName}"></div>
```

构建项目，在浏览器中刷新页面，查看源代码，如图3.6所示。

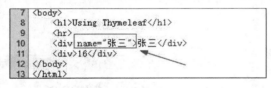 

图 3.6　name 属性值被修改

可以看到浏览器接收到的<div>中增加了name属性，且值为"张三"。这是属性th:name的作用，将元素对应的属性修改为指定值。

其他的属性，如id、class、value等，都可以通过th:加属性名来指定修改。还有更多的用法可以在Thymeleaf官方网站查询。

3. 遍历数组元素

通过th:each可以遍历数组对象。下面在HTML文件中添加一个无序列表来展示其使用。仍在thymeleaf.html文件中添加代码，添加到body内部的最后，代码如下：

```
<ul>
    <li th:each="name : ${studentNames}" th:text="${name}"></li>
</ul>
```

然后，在Controller中添加变量studentNames并赋值，添加到map中，代码如下：

```
String[] studentNames= new String[]{"小明", "小强", "小王"};
map.put("studentNames", studentNames);
```

构建项目，在浏览器中刷新页面，效果如图3.7所示。

代码中的th:each会循环生成所在的标签，用th:text来取出循环变量的内容，放入<li></li>内部。

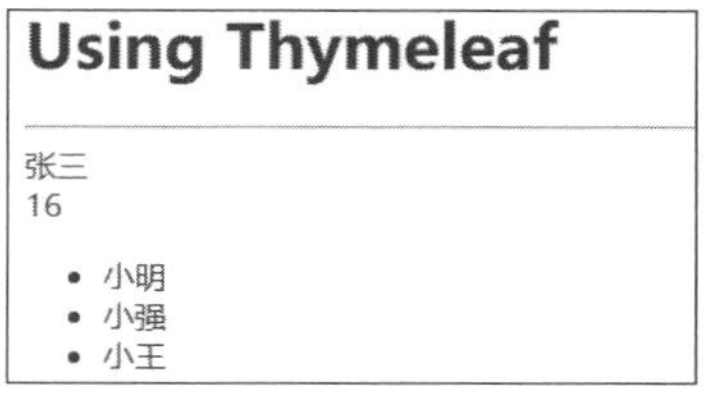

图 3.7 遍历数组生成无序列表

3.3 上 传 文 件

Spring Boot使用Servlet 3的API javax.servlet.http.Part来支持文件上传。Spring Boot在类MultipartAutoConfiguration中定义文件上传组件的自动配置，这个自动配置在使用Spring MVC框架时默认开启。

3.3.1 POM 文件配置

在开启Web模块时，文件上传功能自动开启，所以pom.xml中只需配置spring-boot-starter-web，配置代码如下：

```
<dependency>
  <groupId>org.springframework.boot</groupId>
  <artifactId>spring-boot-starter-web</artifactId>
</dependency>
```

在控制台上，可以通过添加“--debug”启动参数来查看文件上传组件是否已开启。开启时，项目控制台的输出如图3.8所示。

```
   MultipartAutoConfiguration matched:
      - @ConditionalOnClass found required classes 'javax.servlet.Servlet', 'org.springframework.web.multipart.support
.StandardServletMultipartResolver', 'javax.servlet.MultipartConfigElement' (OnClassCondition)
      - found 'session' scope (OnWebApplicationCondition)
      - @ConditionalOnProperty (spring.servlet.multipart.enabled) matched (OnPropertyCondition)

   MultipartAutoConfiguration#multipartConfigElement matched:
      - @ConditionalOnMissingBean (types: javax.servlet.MultipartConfigElement,org.springframework.web.multipart.commons.CommonsMultipartResolver;
SearchStrategy: all) did not find any beans (OnBeanCondition)

   MultipartAutoConfiguration#multipartResolver matched:
      - @ConditionalOnMissingBean (types: org.springframework.web.multipart.MultipartResolver; SearchStrategy: all) did not find any beans
(OnBeanCondition)
```

图 3.8 控制台输出自动配置开启信息

3.3.2 参数设置

在Spring Boot的MultipartProperties类中定义了6个关于文件上传的参数，具体说明如表3.1所示。

表 3.1 文件上传相关参数

名称	默认值	描述
spring.servlet.multipart.enabled	true	是否开启分段上传，默认为true
spring.servlet.multipart.file-size-threshold	0B	文件写入磁盘的阈值
spring.servlet.multipart.location		上传文件的临时目录
spring.servlet.multipart.max-file-size	1MB	上传文件最大大小
spring.servlet.multipart.max-request-size	10MB	文件请求最大大小
spring.servlet.multipart.resolve-lazily	false	是否在文件或参数访问时延迟解析大部分请求

在项目中，可以根据需要在application.yaml中进行配置。在这里，我们使用默认配置。

3.3.3 编写前端页面

为了方便访问，在src/main/resources/static目录下创建HTML文件，文件名为file_upload.html。

在文件中添加一个form表单，用于提交文件，并设置method、action和enctype属性。在该表单内添加两个input，用来选择文件和提交按钮，具体代码如下：

```html
<!DOCTYPE html>
<html lang="en">
<head>
    <meta charset="UTF-8">
    <title>文件上传</title>
</head>
<body>
    <form method="POST" action="/upload" enctype="multipart/form-data">
        <input type="file" name="file" />
        <br/>
        <input type="submit" value="提交" />
    </form>
</body>
</html>
```

这里的method指定了请求类型为POST，请求URL为/upload，下面Controller中的代码要与这些设置相匹配。

3.3.4 编写处理上传请求的 Controller 类

在HTML页面提交会触发请求/upload，因此在服务端需要提供一个处理/upload提交文件的请求，并且将接收到的文件保存到服务器上。在这里将文件保存到D:\files目录下，并且将这个地址作为配置项写到配置文件中。最后，在处理完请求后，请求成功的提示返回给浏览器。具体操作如下：

（1）在src/main/java/com/example/helloworld/controller目录下新创建一个类，命名为FileUploadController.java，代码如下：

```java
package com.example.helloworld.controller;

import org.springframework.beans.factory.annotation.Value;
import org.springframework.stereotype.Controller;
import org.springframework.web.bind.annotation.PostMapping;
import org.springframework.web.bind.annotation.RequestParam;
import org.springframework.web.bind.annotation.ResponseBody;
import org.springframework.web.multipart.MultipartFile;

import java.nio.file.Files;
import java.nio.file.Path;
import java.nio.file.Paths;

@Controller
public class FileUploadController {
    @Value("${save.path}")
    String savePath;
    @ResponseBody
    @PostMapping("/upload")
    public String upload(@RequestParam("file") MultipartFile file) throws Exception{
        String fileSaveName = System.currentTimeMillis()+"-"+file.getOriginalFilename();
        Path saveTo = Paths.get(savePath,fileSaveName);
        Files.write(saveTo, file.getBytes());
        return fileSaveName+",上传成功！";
    }
}
```

（2）在application.yaml中添加save.path的配置，代码如下：

```
save:
  path: D:\files
```

3.3.5　从浏览器上传文件

（1）运行项目，在浏览器中访问http://127.0.0.1:8080/file_upload.html，效果如图3.9所示。

（2）单击"选择文件"按钮，在弹出的窗口中选择文件。这里已经在桌面的file_upload文件夹中放置了一个Spring标识的图片文件，名为spring_logo.png，选中这个文件，如图3.10所示。

图 3.9　文件上传页面

图 3.10　选择文件系统中的图片

单击打开，页面中显示已选择spring_logo.png，如图3.11所示。

（3）单击"提交"按钮，浏览器页面自动刷新，显示"...-spring_logo.png，上传成功！"的字样，如图3.12所示。

这时已经提交成功，并且文件spring_logo.png已经保存到D:\files目录下。打开此目录，可以看到成功保存的文件，如图3.13所示。

  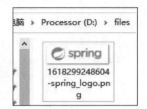

图 3.11　选中图片后的效果　　　图 3.12　上传成功，并显示文件名　　　图 3.13　查看保存的文件

若多次提交，则这个目录下会保存多个前缀不同的-spring_logo.png文件。

3.4　使用定时任务

使用Spring Boot提供的spring-boot-starter-quartz能够快速在项目中集成Quartz Scheduler框架，本节将介绍spring-boot-starter-quartz提供的自动配置和定时任务的使用。

3.4.1　POM 包配置

在pom.xml中添加spring-boot-starter-quartz依赖，代码如下：

```
<dependency>
  <groupId>org.springframework.boot</groupId>
  <artifactId>spring-boot-starter-quartz</artifactId>
</dependency>
```

然后在Maven面板中刷新，使修改生效。使用"--debug"参数启动项目，可以从控制台日志看到Quartz的自动配置已经生效，如图3.14所示。

图 3.14　Quartz 相关日志

3.4.2　对自动配置参数的说明

Spring Boot在QuartzAutoConfiguration.java文件中开启了Quartz的自动配置，以及通过类

QuartzProperties.java提供了自定义参数的变量。这些参数中，spring.quartz.scheduler-name用来指定调取器在容器中的name，spring.quartz.job-store-type用来指定job信息的存储类型。这些参数的作用和默认值可以参考表3.2。

表 3.2 文件上传相关参数

名 称	默 认 值	描 述
spring.quartz.auto-startup	true	在初始化完成后是否启动调度器
spring.quartz.jdbc.comment-prefix	[#, --]	初始化SQL脚本中的单行注释符号
spring.quartz.jdbc.initialize-schema	embedded	启动时进行存储初始化，默认只在使用内存数据库时在启动时进行初始化
spring.quartz.jdbc.schema	classpath:org/quartz/impl/jdbcjobstore/tables_@@platform@@.sql	用来初始化数据库的SQL文件路径
spring.quartz.job-store-type	memory	Quartz job存储类型，memory是在内存中存储，jdbc是使用支持JDBC的数据库存储
spring.quartz.overwrite-existing-jobs	false	由配置文件创建的任务是否覆盖持久化存储中定义的任务
spring.quartz.properties.*		Quartz定时器的其他属性
spring.quartz.scheduler-name	quartzScheduler	调度器在容器中的name
spring.quartz.startup-delay	0s	在初始化完成后，调度程序启动的延时
spring.quartz.wait-for-jobs-to-complete-on-shutdown	false	在关闭程序时是否等待正在运行的任务执行完成

若将spring.quartz.job-store-type指定为jdbc，则Quartz默认使用应用程序的DataSource，通过注解@QuartzDataSource可以为Quartz指定数据源。使用方法很简单，就是在定义数据源的Bean上添加注解@QuartzDataSource。同样，通过注解@QuartzTransactionManager为Quartz指定事务管理器。

3.4.3 编写定时任务代码

创建名为jobs的package，将定时任务相关的代码都放到这个包下。

（1）先创建任务类，编写任务代码。在jobs包下创建类MyJob，继承类QuartzJobBean，并实现其方法。这里注意QuartzJobBean位于org.springframework.scheduling.quartz包下，具体代码如下：

```
package com.example.helloworld.jobs;

import org.quartz.JobExecutionContext;
import org.quartz.JobExecutionException;
import org.springframework.scheduling.quartz.QuartzJobBean;

import java.util.Date;

public class MyJob extends QuartzJobBean {
```

```
        @Override
        protected void executeInternal(JobExecutionContext context) throws
JobExecutionException {
            System.out.println("MyJob:"+new Date());
        }
    }
```

在executeInternal方法中只是做了一个输出,输出当前时间。

(2)在容器中注册JobDetail和Trigger的组件,并将其与MyJob绑定。具体代码如下:

```
package com.example.helloworld.jobs;

import org.quartz.*;
import org.springframework.context.annotation.Bean;
import org.springframework.context.annotation.Configuration;

@Configuration
public class QuartzConfig {
    @Bean
    public JobDetail printTimeJobDetail(){
        return JobBuilder.newJob(MyJob.class)          // 绑定 MyJob
                .withIdentity("MyJobDetail")            // 给该 JobDetail 起一个 id
                .storeDurably()
                .build();
    }
    @Bean
    public Trigger printTimeJobTrigger() {
        CronScheduleBuilder cronScheduleBuilder =
CronScheduleBuilder.cronSchedule("0/1 * * * * ?");
        return (Trigger) TriggerBuilder.newTrigger()
                .forJob("MyJobDetail")                  // 关联上面的 JobDetail
                .withIdentity("quartzTaskService")       // 给 Trigger 起一个名字
                .withSchedule(cronScheduleBuilder)
                .build();
    }
}
```

在上面创建Trigger时用到了CronScheduleBuilder,并使用了cron表达式"0/1 * * * * ?",其含义是每秒钟执行一次,因此在MyJob中的executeInternal方法每秒钟会被执行一次。

3.4.4 测试定时任务执行

通过3.4.1节中的配置和3.4.3节中代码的编写,完成了一个每秒钟执行一次的定时任务,其每秒钟会在控制台输出"MyJob:",后面拼接上当前的时间。

启动项目,控制台输出如图3.15所示。

第 3 章 使用 Spring Boot 进行 Web 开发

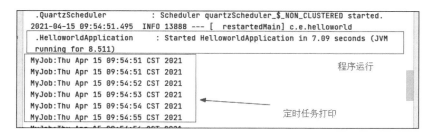

图 3.15 定时任务输出信息

3.5 发 送 邮 件

通过Spring Boot提供的邮件Starter spring-boot-starter-mail能够方便地使用Spring框架的邮件功能。本节将介绍在Spring Boot中使用邮件发送功能。

3.5.1 POM 包配置

在Spring Boot中使用邮件发送功能只需要添加邮件发送的Starter。在pom.xml引入Starter spring-boot-starter-mail依赖，代码如下：

```
<dependency>
  <groupId>org.springframework.boot</groupId>
  <artifactId>spring-boot-starter-mail</artifactId>
</dependency>
```

刷新Maven配置，使变更生效。

3.5.2 在 application.properties 中添加邮箱配置

在通过Starter引入依赖之后，需要通过配置为框架提供必要的参数。

- spring.mail.host：指定 SMTP 服务器的地址，例如使用 QQ 邮箱提供的 SMTP 服务，则应该设置为 smtp.qq.com。
- spring.mail.port：指定 SMTP 服务器的端口，具体信息可以到邮箱服务商的网站上查询。
- spring.mail.username：指定用户名，如果是 QQ 邮箱，则是 QQ 邮箱地址。这个参数也会作为邮件发送人使用。
- spring.mail.port：指定邮箱密码或者授权码。

除了上述必需的参数外，Spring Boot还提供了更多的参数，其信息参见表3.3。

表 3.3 邮件发送相关参数

名 称	默 认 值	描 述
spring.mail.default-encoding UTF-8 Default MimeMessage	UTF-8	默认的MimeMessage编码集

（续表）

名称	默认值	描述
spring.mail.host		SMTP服务器地址，例如smtp.example.com
spring.mail.jndi-name		会话的JNDI名字，优先于会话的其他设置
spring.mail.password		SMTP服务器的登录密码（或授权码）
spring.mail.port		SMTP服务器端口
spring.mail.properties.*		其他JavaMail会话属性
spring.mail.protocol	smtp	SMTP服务器使用的协议
spring.mail.test-connection	false	在程序启动时，是否检测邮件服务器的可用性
spring.mail.username		SMTP服务器用户名
server.spring.sendgrid.api-key		SendGrid API key
spring.sendgrid.proxy.host		SendGrid代理地址
spring.sendgrid.proxy.port		SendGrid代理端口

后面的3.5.4节进行测试时，在application.yaml中所做的配置如下：

```yaml
spring:
  mail:
    host: smtp.qq.com
    port: 587
    username: example@qq.com
    password: examplecode
```

其中的username和password做了脱敏处理，并非真实账号和密码。

3.5.3 编写邮件Service类对框架再封装

在service包下创建类MailService，添加String类型的属性fromAddress，为其注入spring.mail.username属性值；添加MailSender类型的属性mailSender，从容器中注入；最后创建方法sendMail，封装发送邮件的功能。具体代码如下：

```java
package com.example.helloworld.service;

import org.springframework.beans.factory.annotation.Autowired;
import org.springframework.beans.factory.annotation.Value;
import org.springframework.mail.MailException;
import org.springframework.mail.MailSender;
import org.springframework.mail.SimpleMailMessage;
import org.springframework.stereotype.Service;

@Service
public class MailService {
    @Value("${spring.mail.username}")
    private String fromAddress;
    @Autowired
    private MailSender mailSender;

    public String sendMail(String toAddress, String subject, String content){
```

```
    SimpleMailMessage msg = new SimpleMailMessage();
    msg.setFrom(fromAddress);
    msg.setTo(toAddress);
    msg.setSubject(subject);
    msg.setText(content);
    try{
        this.mailSender.send(msg);
        return "success";
    }
    catch (MailException ex) {
        ex.printStackTrace();
        return "failure";
    }
  }
}
```

sendMail方法的3个参数分别是收件人邮箱地址、邮件主题和邮件内容,这样在发送邮件时只需要注入MailService类型的Bean,并调用其sendMail方法即可。

3.5.4 编写测试类进行测试

测试代码放在test目录下,这里使用第1章中helloworld项目的测试类（HelloworldApplicationTests.java）。

（1）在HelloworldApplicationTests.java文件中添加用于注入的属性和测试方法,具体代码如下:

```
@Autowired
MailService mailService;
@Test
void testSendMail(){
    String sendTo = "at_ghr@163.com";
    String subject = "测试邮件";
    String content = "这里是邮件内容";
    mailService.sendMail(sendTo, subject, content);
    System.out.println("邮件发送完成！");
}
```

（2）在testSendMail方法中,先创建3个变量,分别对应MailService类的sendMail方法的3个参数,然后调用sendMail方法发送邮件。

（3）执行测试方法,单击方法左侧的运行测试按钮,如图3.16所示。

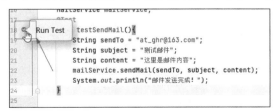

图3.16　启动测试方法

（4）测试程序执行结束，控制台能看到输出"邮件发送完成！"，如图3.17所示。

```
.HelloworldApplicationTests     : Started HelloworldApplicationTests in 5.26
 seconds (JVM running for 8.023)
邮件发送完成！
```

图 3.17　发送邮件控制台打印

登录收件人邮箱，可以看到测试邮件，如图3.18所示。

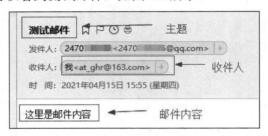

图 3.18　查收邮件

3.6　使用Shiro

Apache Shiro是一个功能强大、使用灵活的开源安全框架，常被用于进行用户认证和权限管理。本节将介绍如何将Shiro、Spring Boot以及模板引擎Thymeleaf整合使用。

3.6.1　基本配置

整合Shiro只需要依赖官方提供的Starter包即可。Starter的信息可以从Shiro官网找到，添加到项目的pom.xml中，具体配置如下：

```xml
<dependency>
  <groupId>org.apache.shiro</groupId>
  <artifactId>shiro-spring-boot-web-starter</artifactId>
  <version>1.7.1</version>
</dependency>
```

由于需要在Thymeleaf模板中使用Shiro方言，因此还需要引入Thymeleaf与Shiro整合的依赖。第三方提供了Thymeleaf与Shiro整合的依赖，可以在Shiro官网的Third-Party Integrations板块找到，具体代码如下：

```xml
<dependency>
  <groupId>com.github.theborakompanioni</groupId>
  <artifactId>thymeleaf-extras-shiro</artifactId>
  <version>2.0.0</version>
</dependency>
```

到这里已经完成整合Shiro的依赖配置，但是为了方便测试，需要预制一些用户和角色信息。Shiro支持从".ini"配置文件中读取用户和角色，借助Shiro的这一特性，我们将测试用的用户和角色信息写入".ini"配置文件中。首先，在resources目录下创建一个名为shiro.ini的文本文件，然后录入用户和角色信息，具体如下：

```
[users]
zhangsan=abcde,admin,order_manager,goods_manager
lisi=123456,order_manager
wangwu=123abc,goods_manager
guest=123123
```

在实际项目中，通常用户和权限（角色）信息都存储在数据库中，因此不会用到".ini"文件。在上面的配置中定义了一个标签"[users]"，其中每一行是一个键值（key-value）对。在这个键值对中，等号左边是用户名，右边第一个是密码，第二个以及之后所有的都是角色。

在上面的设置中共定义了3个用户，分别是zhangsan、lisi和wangwu。以zhangsan为例，他的密码是abcde，他具有两个角色：order_manager和goods_manager。定义的最后一个用户是guest，他的密码是123123，没有分配任何角色。

3.6.2 编写业务逻辑代码和页面

为了便于理解和测试，在整合Shiro之前，先编写一些简单的仿业务逻辑和页面。大致业务逻辑有两个页面：登录页和主页。在登录页输入账号和密码，登录后进入主页。主页显示当前账号以及3个菜单：用户管理、订单管理和商品管理菜单。

（1）创建登录页。在static目录下创建login.html，并创建一个form表单，定义用户名和密码输入框，具体代码如下：

```
<!DOCTYPE html>
<!DOCTYPE html>
<html lang="en">
<head>
    <meta charset="UTF-8">
    <title>登录</title>
</head>
<body>
    <form method="POST" action="/user/login">
        账号：<input type="text" name="username" />
        <br/>
        密码：<input type="password" name="password">
        <br/>
        <input type="submit" value="登录" />
    </form>
</body>
</html>
```

（2）创建主页。在classpath:/templates目录下创建index.html文件，然后创建无序列表来展示3个菜单，使用Thymeleaf语法做账号展示。具体代码如下：

```
<!DOCTYPE html>
<html lang="en">
<head>
```

```html
        <meta charset="UTF-8">
        <title>系统主页</title>
</head>
<body>
        <h2>你好，[[${username}]]！</h2>
        <hr>
        <ul>
            <li>
                <h3><a href="">用户管理</a></h3>
            </li>
            <li>
                <h3><a href="">订单管理</a></h3>
            </li>
            <li>
                <h3><a href="">商品管理</a></h3>
            </li>
        </ul>
</body>
</html>
```

只有页面还不够，还需要两个接口，一个实现登录逻辑，接收前台传过来的用户名和密码，并在登录成功后跳转到系统主页；另一个接口承接主页（/user/index.html）的访问。

在Controller目录下创建一个文件UserController.java，并在其中完成这两个接口。具体代码如下：

```java
package com.example.helloworld.controller;

import org.springframework.stereotype.Controller;
import org.springframework.web.bind.annotation.RequestMapping;

@Controller
@RequestMapping("user")
public class UserController {
    @RequestMapping("login")
    public String login(String username, String password){
        System.out.println("用户登录："+username);
        if(true){   // TODO 实现用户认证
            return "redirect:/user/index.html";
        }
        return "redirect:/login.html";          // 认证失败，跳转回登录页
    }
    @RequestMapping("index.html")
    public String loginPage(){                  // 系统主页
        return "index.html";
    }
}
```

接下来对两个页面存放路径进行说明。login.html中没有使用Thymeleaf模板，因此直接放到了static目录下，可以在URL中直接访问。而index.html使用到了Thymeleaf语法，而且在下面引入Shiro时需要权限相关标签，这些是要被解析的，因此存放到了Thymeleaf可识别的目录templates中，同时需要在Controller中映射请求来访问。

这里为了查看页面效果，先临时注释pom.xml中两个Shiro相关的依赖，然后启动项目，访问http://127.0.0.1:8080/login.html，浏览器显示如图3.19所示。

输入用户名和密码，单击"登录"按钮，会跳转到http://127.0.0.1:8080/user/index.html，效果如图3.20所示。注意，由于未设置账号，"你好，！"中没有显示账号信息。

图3.19　登录页

图3.20　系统主页

看过效果之后，把pom.xml中的注释打开，以便接下来使用Shiro。

3.6.3　在代码中引入 Shiro

在Web项目中引入Shiro需要配置ShiroFilter，用来拦截所有需要认证的请求。通常除了登录页、登录请求以及部分静态资源外，其余的所有请求都需要用户在访问前先登录，基本的请求拦截模型如图3.21所示。

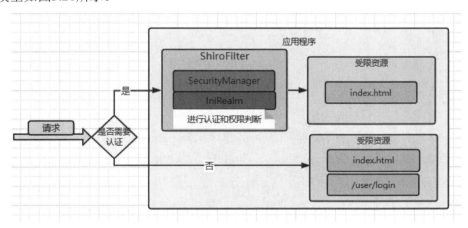

图3.21　拦截请求的大致思路

（1）首先，在容器中创建ShiroFilter、SecurityManager和IniRealm对象，并将IniRealm和SecurityManager绑定，SecurityManager与ShiroFilter绑定。具体操作是创建一个com.example.helloworld.config包，然后在包中创建一个配置类ShiroConfig.java。创建ShiroFilter、SecurityManager和IniRealm这3个代码Bean，具体代码如下：

```
package com.example.helloworld.config;

import org.apache.shiro.spring.web.ShiroFilterFactoryBean;
import org.apache.shiro.web.mgt.DefaultWebSecurityManager;
```

```java
import org.springframework.context.annotation.Bean;
import org.springframework.context.annotation.Configuration;
@Configuration
public class ShiroConfig {
    @Bean
    // 使用 Shiro starter 提供的工厂类创建 ShiroFilter 对象
    public ShiroFilterFactoryBean shiroFilterFactoryBean(DefaultWebSecurityManager defaultWebSecurityManager){
        ShiroFilterFactoryBean shiroFilterFactoryBean = new ShiroFilterFactoryBean();
        // 将 SecurityManager 与 ShiroFilter 绑定
        shiroFilterFactoryBean.setSecurityManager(defaultWebSecurityManager);
        return shiroFilterFactoryBean;
    }
    @Bean
    // 在 Web 环境中使用 DefaultWebSecurityManager 创建对象
    public DefaultWebSecurityManager defaultWebSecurityManager(IniRealm iniRealm){
        DefaultWebSecurityManager defaultWebSecurityManager = new DefaultWebSecurityManager();
        // 将 IniRealm 绑定到 DefaultWebSecurityManager
        defaultWebSecurityManager.setRealm(iniRealm);
        return defaultWebSecurityManager;
    }
    @Bean
    public IniRealm iniRealm(){
        // 从配置文件 shiro.ini 中读取用户和角色信息
        return new IniRealm("classpath:shiro.ini");
    }
}
```

到这里，ShiroFilter默认会拦截所有请求，并重定向到login.jsp。我们的登录页是login.html，所以还需要配置登录页为login.html，并且登录请求为/user/login。这需要在方法shiroFilterFactoryBean返回之前添加如下代码：

```java
// 设置登录页为 login.html，在未登录访问时，会自动重定向到此地址
shiroFilterFactoryBean.setLoginUrl("/login.html");
```

但是地址"/login.html"以及登录请求地址"/user/login"仍会被ShiroFilter拦截，所以需要为这两个请求配置可以匿名访问的过滤器AnonymousFilter，具体代码如下：

```java
Map<String,String> map = new LinkedHashMap<>();
map.put("/login.html","anon");
map.put("/user/login","anon");
// 配置认证和授权规则
shiroFilterFactoryBean.setFilterChainDefinitionMap(map);
```

（2）至此，已经完成Shiro的配置工作。下面开始在业务逻辑代码中引入Shiro。

在UserController的login方法中使用Shiro的API完成登录功能，具体代码如下：

```
public String login(String username, String password){
    System.out.println("用户登录："+username);
    // 获取subject对象
    Subject subject = SecurityUtils.getSubject();
    try{
        // 实现登录
        subject.login(new UsernamePasswordToken(username, password));
        // 登录成功，重定向到主页
        return "redirect:/user/index.html";
    } catch (UnknownAccountException e) {
        System.out.println("用户名错误！");
    } catch (IncorrectCredentialsException e) {
        System.out.println("密码错误！");
    } catch (Exception e){
        System.out.println("未知异常！");
    }
    return "redirect:/login.html";
}
```

需要在index.html页面中获取账号信息，因此需要在loginPage方法中将账号信息设置到请求域中。在方法参数中定义一个Map<String, Object>类型变量，该变量会默认放到请求域中，在该map对象中设置账号信息，具体代码如下：

```
@RequestMapping("index.html")
public String loginPage(Map<String, Object> map){ // 系统主页
    // 获取subject对象
    Subject subject = SecurityUtils.getSubject();
    // 获取主身份信息，在这里也就是账号username
    String username = (String)subject.getPrincipal();
    map.put("username", username);
    return "index.html";
}
```

（3）在index.html页面中进行权限管理。

使用shiro:hasRole来指定需要某一个权限才能访问的HTML资源。在3.6.1节中已经引入了Shiro和Thymeleaf整合的依赖，但是要使Shiro方言生效，还需要在容器中添加ShiroDialect组件。不妨在ShiroConfig.java类中创建，具体代码如下：

```
@Bean(name = "shiroDialect")
public ShiroDialect shiroDialect(){
    return new ShiroDialect();
}
```

在此处的业务逻辑中，设置"用户管理"只能由admin角色查看，"订单管理"只能由order_manager角色查看，"商品管理"只能由goods_manager查看。具体代码如下：

```
<ul>
    <li shiro:hasRole="admin">
        <h3><a href="">用户管理</a></h3>
```

```
        </li>
        <li shiro:hasRole="order_manager">
            <h3><a href="">订单管理</a></h3>
        </li>
        <li shiro:hasRole="goods_manager">
            <h3><a href="">商品管理</a></h3>
        </li>
    </ul>
```

至此，已完成所有代码。在下一节中观察引入Shiro后的效果。

3.6.4 测试用户认证和权限管理的效果

启动项目，在浏览器中访问之前，先依据在配置文件shiro.ini中定义的用户和角色分析展示效果。

在shiro.ini中共定义了4个账号，分别是zhangsan、lisi、wangwu和guest。这4个账号有各自不同的密码，登录时输入不存在的账号，后台会打印用户名错误；输入错误的密码，后台会打印密码错误，同时浏览器跳转回登录页。至于角色，在4个账号中，用户zhangsan拥有所有角色，因此能看到所有菜单，而lisi和wangwu只能分别看到其角色对应的菜单。最后的账号guest由于没有任何角色，因此看不到任何菜单。下面测试效果。

（1）输入不存在的账号。在浏览器中访问http://127.0.0.1:8080/login.html，输入账号zhangsan1，密码123456，单击"登录"按钮，浏览器跳转回登录页，控制台输出如图3.22所示。

（2）输入错误的密码。输入账号zhangsan，密码12345，单击"登录"按钮，浏览器跳转回登录页，控制台输出如图3.23所示。

图 3.22　用户名错误

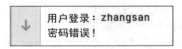

图 3.23　密码错误

（3）使用zhangsan登录，密码为abcde，成功登录，index.html页面展示zhangsan账号，显示3个菜单，如图3.24所示。

（4）使用lisi登录，密码为123456，成功登录，index.html页面展示lisi账号，仅显示菜单订单管理，如图3.25所示。

图 3.24　登录后主页的效果

图 3.25　lisi 的主页

（5）使用wangwu登录，密码为123abc，成功登录，index.html页面展示wangwu账号，仅显示菜单商品管理，如图3.26所示。

图 3.26 wangwu 的主页

（6）最后，使用guest登录，密码为123123，index.html页面展示wangwu账号，但没有任何菜单，如图3.27所示。

图 3.27 guest 的主页

3.7 实战——开发一个简单的Restful API网关

本节将开发一个简单的网关，实现请求转发功能。从业务逻辑上讲，继续在helloworld项目中将/account/**的请求转发到/user/**下。具体来讲，需要接收所有以/account开头的请求，从请求中获取请求方式、HTTP头和请求体的信息并进行封装，然后/user发出请求，最后获取响应数据返回给/account的请求。具体步骤如下：

（1）在com.example.helloworld.controller包下创建类GatewayController.java，并使用注解@RestController和@RequestMapping("account")来接收/account开头的请求。代码如下：

```
package com.example.helloworld.controller;

import org.springframework.web.bind.annotation.RequestMapping;
import org.springframework.web.bind.annotation.RestController;

@RestController
@RequestMapping("account")
public class GatewayController {
}
```

（2）在GatewayController中创建方法catchRequest，返回值类型为ResponseEntity，并且创建参数为HttpServletRequest和HttpServletResponse，方便获取请求和返回响应数据。还要使用注解RequestMapping来指定所有请求类型和URL的请求。代码如下：

```
@RequestMapping(value = "/**", method = {RequestMethod.GET, RequestMethod.POST, RequestMethod.PUT, RequestMethod.DELETE})
public ResponseEntity catchRequest(HttpServletRequest request, HttpServletResponse response) {
    return null;
}
```

（3）接下来创建类DelegateService.java，在这个类中实现封装请求、转发请求的功能，使用到了Spring框架的RestTemplate API。具体DelegateService.java类的代码实现如下，其实现过程在注释中做了说明：

```
package com.example.helloworld.service;

import org.springframework.http.*;
import org.springframework.stereotype.Service;
import org.springframework.util.MultiValueMap;
import org.springframework.util.StreamUtils;
import org.springframework.web.client.RestTemplate;
import javax.servlet.http.HttpServletRequest;
import javax.servlet.http.HttpServletResponse;
import java.io.IOException;
import java.io.InputStream;
import java.net.URI;
import java.net.URISyntaxException;
import java.util.Collections;
import java.util.List;

@Service
public class DelegateService {
    /**
     * 转发入口方法
     * @param request    原请求对象
     * @param response   原响应对象
     * @param routeUrl   目标地址
     * @param prefix     被替换的前缀
     * @return
     */
    public ResponseEntity<String> redirect(HttpServletRequest request, HttpServletResponse response, String routeUrl, String prefix) {
        try {
            // 创建转发目标请求 URL
            String redirectUrl = createRedictUrl(request,routeUrl, prefix);
            RequestEntity requestEntity = createRequestEntity(request, redirectUrl);
            return route(requestEntity);
        } catch (Exception e) {
```

```
                return new ResponseEntity("REDIRECT ERROR",
HttpStatus.INTERNAL_SERVER_ERROR);
            }
        }
        /**
         * 从请求中获取请求路径和参数,封装到被代理的请求路径上
         */
        private String createRedictUrl(HttpServletRequest request, String routeUrl,
String prefix) {
            String queryString = request.getQueryString();
            return routeUrl + request.getRequestURI().replace(prefix, "") +
                    (queryString != null ? "?" + queryString : "");
        }
        /**
         * 从源请求中获取请求类型、HTTP 头和请求体信息,以及目标 URL,封装成 RequestEntity
返回
         * @param request 源请求对象
         * @param url 目标 URL
         * @return
         * @throws URISyntaxException
         * @throws IOException
         */
        private RequestEntity createRequestEntity(HttpServletRequest request,
String url) throws URISyntaxException, IOException {
            String method = request.getMethod();
            HttpMethod httpMethod = HttpMethod.resolve(method);
            MultiValueMap<String, String> headers = parseRequestHeader(request);
            byte[] body = parseRequestBody(request);
            return new RequestEntity<>(body, headers, httpMethod, new URI(url));
        }
        /**
         * 使用 RestTemplate 发送请求,并返回响应结果
         * @param requestEntity
         * @return
         */
        private ResponseEntity<String> route(RequestEntity requestEntity) {
            RestTemplate restTemplate = new RestTemplate();
            restTemplate.getMessageConverters().set(1, new
StringHttpMessageConverter(StandardCharsets.UTF_8));
            return restTemplate.exchange(requestEntity, String.class);
        }
        /**
         * 从请求对象中获取请求体
         * @param request
         * @return
         * @throws IOException
         */
```

```java
    private byte[] parseRequestBody(HttpServletRequest request) throws 
IOException {
        InputStream inputStream = request.getInputStream();
        return StreamUtils.copyToByteArray(inputStream);
    }
    /**
     * 从请求对象中获取 HTTP 头信息,并封装到 Map 中
     * @param request
     * @return
     */
    private MultiValueMap<String, String> 
parseRequestHeader(HttpServletRequest request) {
        HttpHeaders httpHeaders = new HttpHeaders();
        List<String> headerNames = 
Collections.list(request.getHeaderNames());
        for (String headerName : headerNames) {
            List<String> headerValues = 
Collections.list(request.getHeaders(headerName));
            for (String headerValue : headerValues) {
                httpHeaders.add(headerName, headerValue);
            }
        }
        return httpHeaders;
    }
}
```

(4) 在GatewayController中注入DelegateService, 并完善catchRequest方法, 实现转发功能。GatewayController.java的完整代码如下:

```java
package com.example.helloworld.controller;

import com.example.helloworld.service.DelegateService;
import org.springframework.beans.factory.annotation.Autowired;
import org.springframework.http.ResponseEntity;
import org.springframework.web.bind.annotation.RequestMapping;
import org.springframework.web.bind.annotation.RequestMethod;
import org.springframework.web.bind.annotation.RestController;

import javax.servlet.http.HttpServletRequest;
import javax.servlet.http.HttpServletResponse;

@RestController
@RequestMapping("account")
public class GatewayController {
    /**
     * 指定要被替换的前缀
     */
    public static final String prefix = "/account/";
    /**
     * 被代理的请求地址, 这里还是访问当前项目, 效果是将"account"换成了"user"
```

```
    */
    public static final String routeUrl = "http://127.0.0.1:8080/user";
    @Autowired
    DelegateService delegateService;

    @RequestMapping(value = "/**", method = {RequestMethod.GET,
RequestMethod.POST, RequestMethod.PUT, RequestMethod.DELETE})
    public ResponseEntity catchRequest(HttpServletRequest request,
HttpServletResponse response) {
        return delegateService.redirect(request, response, routeUrl,
"account");
    }
}
```

（5）上面完成了网关的全部开发，下面测试网关。

依据设计的逻辑，当在浏览器访问http://127.0.0.1:8080/account/login时，我们的网关会将请求转发到http://127.0.0.1:8080/user/login，但是由于没有传递username和password，后台获取到的username为null，输出"用户名错误"，如图3.28所示。

图 3.28　用户名为 null

与此同时，浏览器展现登录页面，如图3.29所示。

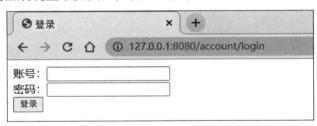

图 3.29　/account/login 跳转到登录页

由此，可以说明确实访问了/user/login，网关转发成功。

第 4 章

使用Spring Boot进行数据库开发

在项目中几乎必不可少使用数据库对业务数据进行管理,包括常见的增删改查等操作。本章提到的数据库指SQL Database。本章将以MySQL为例说明如何配置数据源以及整合常见的持久层框架。

本章主要涉及的知识点有:

- 如何配置一个数据源。
- 如何使用JdbcTemplate。
- JPA是什么,如何使用Spring Data JPA。
- MyBatis是什么,如何使用MyBatis。
- MyBatis如何与Spring Boot项目整合。

4.1 配置数据源

数据源(DataSource)的作用是连接数据库和管理连接池,其性能将直接影响项目的性能。目前有众多数据源产品,如C3P0、DBCP、HikariCP、Druid等。本节将介绍如何在Spring Boot项目中配置我们需要的数据源。

4.1.1 启动默认数据源

默认数据源的数据库连接信息在配置文件的spring.datasource.*项中进行配置。例如,我们可以在application.properties中采用如下形式进行声明:

```
spring.datasource.url=jdbc:mysql://127.0.0.1:3306/db
spring.datasource.username=dbusername
spring.datasource.password=123456
```

或者在application.yaml中进行配置:

```yaml
spring:
  datasource:
    url: "jdbc:mysql://127.0.0.1:3306/db"
    username: "dbusername"
    password: "123456"
```

数据源的类型可以由spring.datasource.type来指定（DataSource实现类的完全限定名），如果没有指定，则Spring Boot选择默认的数据源，其选取规则如下：

（1）首先，如果HikariCP可用，则优先选择HikariCP。
（2）其次，检查Tomcat的数据库连接池，若可用，则使用。
（3）再次，检查DBCP2是否可用，若可用，则使用。
（4）最后，检查Oracle UCP是否可用。

其过程图4.1所示。

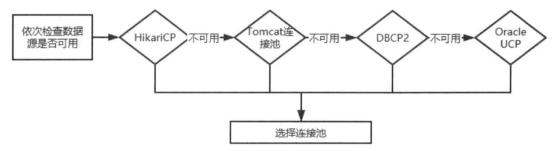

图 4.1　Spring Boot 默认选择数据源的规则

此外，可以使用spring.datasource.driver-class-name来设置JDBC驱动类。由于Spring Boot根据spring.datasource.url来自动选择驱动类，因此通常我们无须显式指定。

4.1.2　配置自定义数据源

在Spring Boot中配置自定义数据源，可以通过注解@Bean将自定义的DataSource注册到Spring容器中。例如，创建config包，在包下创建类CustomDataSourceConfig.java，并使用dataSource()方法注册DataSource实例，代码如下：

```
@Component
public class CustomDataSourceConfig {
    @Bean
    @ConfigurationProperties(prefix="custom.datasource")
    public DataSource dataSource() {
        return new CustomDataSource();
    }
}
```

其中@ConfigurationProperties(prefix="custom.datasource")指定配置文件的前缀为custom.datasource，在application.yaml配置文件中修改如下：

```yaml
custom:
  datasource:
    url: "jdbc:mysql://127.0.0.1:3306/db"
    username: "dbusername"
    password: "123456"
```

其中，custom.datasource.*中的配置，如url、username和password，对应类CustomDataSource中定义的属性。

4.2 使用JdbcTemplate操作数据库

在Spring Boot中开启访问数据库，首先要添加JDBC的依赖，在pom.xml文件中添加如下配置：

```xml
<dependency>
    <groupId>org.springframework.boot</groupId>
    <artifactId>spring-boot-starter-jdbc</artifactId>
</dependency>
```

完成以上配置后，接下来通过在Test中调用jdbcTemplate查询来完成对数据库的操作。

（1）在数据库db中创建user，并在其中插入两条数据：

```sql
CREATE TABLE user (
  id int NOT NULL,
  name varchar(255) DEFAULT NULL,
  age int DEFAULT NULL,
  address varchar(255) DEFAULT NULL,
  PRIMARY KEY (id)
) ENGINE=InnoDB DEFAULT CHARSET=utf8mb4 COLLATE=utf8mb4_0900_ai_ci;
INSERT INTO user(id, name, age, address) VALUES (1, '张三', 18, '新华街');
INSERT INTO user(id, name, age, address) VALUES (2, '李四', 22, '康庄路');
```

（2）在Spring Boot测试类DemoApplicationTests中添加测试代码：

```java
@Autowired
JdbcTemplate jdbcTemplate;
@Test
void testTemplate() {
    Integer count = .queryForObject("select count(id) from user", Integer.class);
    System.out.println("users 表中共有数据条数："+count);
}
```

（3）运行该测试方法，控制台输出"users表中共有数据条数：2"，表示以上配置正确无误。同时观察日志，发现HikariDataSource相关日志，如图4.2所示，表示当前Spring Boot默认使用的是HikariDataSource数据源。

图4.2 测试配置默认数据源控制台输出日志

至此，便能够通过JdbcTemplate访问数据库，对数据进行增删改查等操作。JdbcTemplate提供了一系列接口，具体可以参考官方提供的API文档。

4.3 使用Spring Data JPA（Hibernate）操作数据

使用Spring Data JPA能够实现基于JPA的数据存储操作。本节将介绍Spring Data JPA在Spring Boot项目中的使用。

4.3.1 基础知识

要理解Spring Data JPA，需要对JPA、Hibernate JPA和Spring Data JPA进行区分。

（1）JPA是Java Persistence API的简称，中文可以翻译为Java持久层API，最早是由sun公司提供的ORM（Object Relational Mapping）对象关系映射的解决方案，统一了使用ORM访问持久层的方式。也可以理解成JPA是没有提供具体实现的一套ORM标准。

（2）Hibernate ORM是一套独立的ORM框架，它提供实现ORM的一套自有（Native）API。同时，Hibernate在它自有的API之外，为方便在支持JPA的环境中使用Hibernate，提供了一套对JPA标准的实现，这套基于JPA实现的API也就是Hibernate JPA。就功能来讲，JPA是Hibernate功能的一个子集。JPA和Hibernate的关系如图4.3所示，有点类似JDBC和JDBC驱动的关系。

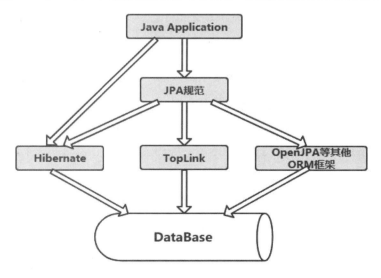

图4.3 JPA 规范与 ORM 框架关系示意图

（3）Spring Data JPA也是一套ORM标准，是对JPA标准的增强，它最引人注目的特性是能够在运行时从存储库接口自动创建存储库实现。Spring Data JPA也是没有自己的实现，在Spring Data JPA的Spring Boot Starter整合时默认使用Hibernate作为实现。

4.3.2 依赖管理和配置信息

只需要引入spring-boot-starter-data-jpa依赖即可开启Spring Boot JPA，在pom.xml中配置如下：

```xml
<dependency>
  <groupId>org.springframework.boot</groupId>
  <artifactId>spring-boot-starter-data-jpa</artifactId>
</dependency>
```

在控制台上，可以通过添加"--debug"启动参数来查看文件上传组件是否已开启。开启时，启动项目控制台输出如图4.4所示。

```
JpaRepositoriesAutoConfiguration matched:
    - @ConditionalOnClass found required class 'org.springframework.data.jpa
.repository.JpaRepository' (OnClassCondition)
    - @ConditionalOnProperty (spring.data.jpa.repositories.enabled=true) matched
(OnPropertyCondition)
    - @ConditionalOnBean (types: javax.sql.DataSource; SearchStrategy: all) found
bean 'dataSource'; @ConditionalOnMissingBean (types: org.springframework.data.jpa
.repository.support.JpaRepositoryFactoryBean,org.springframework.data.jpa.repository
.config.JpaRepositoryConfigExtension; SearchStrategy: all) did not find any beans
(OnBeanCondition)
```

图 4.4 控制台输出自动配置开启信息

并且能看到对Hibernate JPA的自动配置，如图4.5所示。

```
HibernateJpaAutoConfiguration matched:
    - @ConditionalOnClass found required classes 'org.springframework.orm.jpa
.LocalContainerEntityManagerFactoryBean', 'javax.persistence.EntityManager',
'org.hibernate.engine.spi.SessionImplementor' (OnClassCondition)

HibernateJpaConfiguration matched:
    - @ConditionalOnSingleCandidate (types: javax.sql.DataSource; SearchStrategy:
all) found a primary bean from beans 'dataSource' (OnBeanCondition)
```

图 4.5 Hibernate JPA 自动配置信息

对于JPA和Hibernate的配置，Spring Boot提供了配置参数，具体在表4.1中进行说明。

表 4.1 Spring Data JPA 参数设置

名 称	默 认 值	描 述
spring.data.jpa.repositories.bootstrap-mode	default	JPA Repository的启动模式（Bootstrap Mode）
spring.data.jpa.repositories.enabled	true	是否开启JPA Repository
spring.data.ldap.repositories.enabled	true	是否开启LDAP Repository
spring.jpa.database		要操作的数据库，默认自动检测。也可以使用databasePlatform属性进行设置

（续表）

名称	默认值	描述
spring.jpa.database-platform		要操作的数据库名称，默认自动检测。也可以通过database属性进行设置
spring.jpa.generate-ddl	false	是否在启动时执行初始化语句
spring.jpa.hibernate.ddl-auto		DDL模式。这个属性值会被设置到Hibernate的hibernate.hbm2ddl.auto属性中。当使用嵌入式数据库并且没有检测到模式管理器时，默认值是create-drop；否则，默认值是none
spring.jpa.hibernate.naming.implicit-strategy		隐式命名策略的完全限定名
spring.jpa.hibernate.naming.physical-strategy		物理命名策略的完全限定名
spring.jpa.hibernate.use-new-id-generator-mappings		是否为AUTO，TABLE和SEQUENCE使用Hibernate新款的ID生成器
spring.jpa.mapping-resources		映射资源。等价于persistence.xml中的mapping-file
spring.jpa.open-in-view	true	是否注册OpenEntityManagerInViewInterceptor。用来将JPA EntityManager绑定到用于整个请求处理的线程
spring.jpa.properties.*		其他JPA提供者的属性。其设置内容将设置给JPA提供组件
spring.jpa.show-sql	false	是否在日志中打印执行的SQL语句

关于其中的spring.jpa.properties.*的使用，举一个例子，如果需要设置的Hibernate的参数如下：

```
hibernate.globally_quoted_identifiers=true
```

那么在application.properties中的设置需要加上前缀，即spring.jpa.properties.，如下：

```
spring.jpa.properties.hibernate.globally_quoted_identifiers=true
```

这样实际上hibernate.globally_quoted_identifiers就会被设置给Hibernate。

4.3.3 使用 Spring Data JPA 进行开发

使用Spring Data JPA进行开发主要是使用接口和注解来实现的。下面将通过具体注解和接口的使用来学习Spring Data JPA。

1. 关联实体类和数据库表

使用JPA提供的注解可以将实体类和数据库表关联起来，比如注解@Table将某个实体类和某个数据库表关联，@Column将字段和数据库表的一列关联。更多常用的注解列举在表4.2中。

表 4.2 用于关联实体类和数据库表的注解

名　　称	标注位置	使用及注解属性说明
@Entity	类	指明标注类为实体类
@Table	类	用于指定与实体类映射的数据库表的名称。 元数据如下： • name：表名。 • catalog：对应关系数据库中的catalog。 • schema：对应关系数据库中的schema。 • UniqueConstraints：UniqueConstraint类型数组，指定需要创建唯一约束的列。 • indexes：指定表的索引
@Id	字段、方法	映射到数据库表的主键的属性，一个实体只能有一个属性被映射为主键
@GeneratedValue	字段、方法	配合注解@Id来使用，用来标注主键的生成策略。 元数据如下： • strategy：表示主键生成策略，有AUTO、INDENTITY、SEQUENCE和TABLE 4种。默认值为AUTO，表示让ORM框架根据数据库类型自动选择。 • generator：表示主键生成器的名称，这个属性通常和ORM框架相关
@Basic	字段、方法	指定字段抓取策略和是否允许为空。 元数据如下： • fetch：抓取策略，可选LAZY或EAGER，默认值为EAGER。 • optional：指定字段是否允许为null，默认值为true
@Column	字段、方法	指定字段的详细定义。 元数据如下： • name：字段的名称，默认与属性名称一致。 • unique：是否唯一，默认值为false。 • nullable：是否允许为空，默认值为true。 • insertable：是否被包含在ORM框架生成的SQL INSERT语句中，默认值为true。 • updatable：是否被包含在ORM框架生成的SQL UPDATE语句中，默认值为true。 • columnDefinition：用于DDL语句中指定字段的类型，默认ORM框架采用推断的类型。 • table：列所属的表名，默认列属于主表。 • length：字段的长度，仅对String类型的字段有效。 • precision：字段的精确度，仅对数值类型有效。如果生成DDL语句，则必须指定。 • scale：数值类型列的大小范围，仅对数值类型有效
@Transient	字段、方法	被标注的字段被ORM框架忽略与数据库中的字段映射

（续表）

名称	标注位置	使用及注解属性说明
@OneToOne	字段、方法	指定一对一关联类型的与另一个实体的关联字段。通常不需要指定关联的数据库表，可以由字段类型推断出来。 元数据属性说明： - targetEntity：指定关联类型，默认为字段的实体类型。 - cascade：定义级联操作的类型，数组类型，有6个可选：ALL、PERSIST、MERGE、REMOVE、REFRESH和DETACH。 - fetch：关联是应该惰性加载还是必须急切获取。类似于@Basic的fetch，可选LAZY或EAGER，默认值为EAGER。 - optional：关联是否是可选的。默认值为true，如果设置为false，则非空关系必须始终存在。 - mappedBy：对于双边的关联，这个属性用于被关联方，指定的是源官方的字段名。 - orphanRemoval：是否将删除操作应用于已从关系中删除的实体，并将删除操作级联到这些实体。默认值为False
@ManyToOne	字段、方法	指定多对一的映射，通常不需要显式地指定目标实体，因为它通常可以从被引用的对象的类型推断出来。 元数据如下： - targetEntity：关联到的实体类型，默认是字段类型。 - cascade：定义级联操作的类型，数组类型，有6个可选：ALL、PERSIST、MERGE、REMOVE、REFRESH和DETACH。 - fetch：关联是应该惰性加载还是必须急切获取。类似于@Basic的fetch，可选LAZY或EAGER，默认值为EAGER。 - optional：关联是否是可选的。默认值为true，如果设置为false，则非空关系必须始终存在
@OneToMany	字段、方法	指定一对多的关联关系，该属性应该为集合类型，在数据库中并没有实际字段。 元数据如下： - targetEntity：关联到的实体类型，默认是字段类型。 - cascade：定义级联操作的类型，数组类型，有6个可选：ALL、PERSIST、MERGE、REMOVE、REFRESH和DETACH。 - fetch：关联是应该惰性加载还是必须急切获取。类似于@Basic的fetch，可选LAZY或EAGER，默认值为EAGER。 - mappedBy：对于双边的关联，这个属性用于被关联方，指定的是源官方的字段名。 - orphanRemoval：是否将删除操作应用于已从关系中删除的实体，并将删除操作级联到这些实体。默认值是false

（续表）

名称	标注位置	使用及注解属性说明
@JoinColumn	字段、方法	用来指定关联实体类的外键字段，外键可以在当前实体类中，也可以在关联表中。 元数据如下： • name：指定外键列名，仅在使用单个连接列时适用。 • referencedColumnName：指定被关联表的列名，仅在使用单个连接列时适用。 • unique：属性是否为唯一键。这是表级的uniqueconstraint注释的快捷方式，当唯一键约束只有一个字段时非常有用。对于与作为外键一部分的主键相对应的连接列，没有必要显式地指定此值。 • nullable：外键列是否允许为空，默认值为true。 • insertable：是否被包含在ORM框架生成的SQL INSERT语句中，默认值为true。 • updatable：是否被包含在ORM框架生成的SQL UPDATE语句中，默认值为true。 • columnDefinition：用于DDL语句中指定字段的类型，默认ORM框架采用推断的类型。 • table：列所属的表名，默认列属于主表。 • foreignKey：指定是否使用外键约束，或者使用持久化框架提供的方案。默认使用持久化框架提供的方案
@JoinColumns	字段、方法	存在多个关联关系时，用于指定多个@JoinColumn。 元数据如下： • value：@JoinColumn的数组类型，用于指定多个@JoinColumn。 • foreignKey：指定是否使用外键约束，或者使用持久化框架提供的方案。默认使用持久化框架提供的方案
@Lob	字段、方法	应该用于标注数据量大的类型，比如用于标注字段类型为Clob和Blob类型。基于字符的类型如String，默认会被存储为Clob。 仅做标注使用，没有元数据属性
@ManyToMany	字段、方法	指定多对多的关联关系。多对多关联是两个一对多关联，但是通过@ManyToMany描述，中间表由ORM框架自动处理。 元数据如下： • targetEntity：指定关联的实体类类型，默认为集合类指定的泛型。 • cascade：定义级联操作的类型，数组类型，有6个可选：ALL、PERSIST、MERGE、REMOVE、REFRESH和DETACH。对于Map类型，级联操作对应Map的value。 • fetch：关联是应该惰性加载还是必须急切获取。类似于@Basic的fetch，可选LAZY或EAGER，默认值为EAGER。 • mappedBy：对于双边的关联，这个属性用于被关联方，指定的是源官方的字段名

（续表）

名称	标注位置	使用及注解属性说明
@JoinTable	方法、字段	用于在many-to-many关系的所有者一边定义。如果没有定义@JoinTable，则使用@JoinTable的默认值。 元数据如下： • name：关联的中间表表名，默认值为两个关联的主实体表的连接名，由下画线分隔。 • catalog：数据库关联表的catalog。 • schema：数据库关联表的schema。 • joinColumns：指定用于关联的外键列，这个外键列属于当前实体类。 • inverseJoinColumns：指定用于关联的外键列，这个外键列属于被关联实体类。 • foreignKey：指定外键约束的生成方式，指定是否使用外键约束，或者使用持久化框架提供的方案。默认使用持久化框架提供的方案。 • inverseForeignKey：指定被关联表的关联字段外键约束的生成方式，指定是否使用外键约束，或者使用持久化框架提供的方案。默认使用持久化框架提供的方案。 • UniqueConstraints：UniqueConstraint类型数组，指定需要创建唯一约束的列。 • indexes：指定表的索引

下面的示例是一对一关联关系的使用。

【示例4.1】 学生和联系方式的一对一关系

学生实体类表示学生的基本信息，联系方式对应学生监护人、家庭住址、联系电话等信息。学生关联一个联系方式，联系方式中关联一个学生，用联系方式的id进行关联，并且关联字段放在学生表中。

学生实体类Student.java代码如下：

```
package com.example.school.pojo;

import javax.persistence.*;

@Entity
@Table(name = "student")
public class Student {
    @Id
    @GeneratedValue(strategy = GenerationType.IDENTITY)    // 设置id生成方式为自增长
    private Integer id;
    @Column
    private String name;
    @Column
    private Integer age;
    @Column
    private String gender;
```

```java
        @OneToOne(cascade = CascadeType.PERSIST)           // 在插入时保持级联操作
        @JoinColumn(name = "contact_id", referencedColumnName = "id")
        private Contact contact;
        public Student() {
        }

        public Student(Integer id, String name, Integer age, String gender, Contact contact) {
            this.id = id;
            this.name = name;
            this.age = age;
            this.gender = gender;
            this.contact = contact;
        }
        //  get、set 方法

        @Override
        public String toString() {
            return "Student{" +
                    "id=" + id +
                    ", name='" + name + '\'' +
                    ", age=" + age +
                    ", gender='" + gender + '\'' +
                    ", contact=" + contact +
                    '}';
        }
    }
```

联系方式实体类Contact.java代码如下：

```java
package com.example.school.pojo;

import javax.persistence.*;

@Entity
@Table(name = "contact")
public class Contact {
    @Id
    @GeneratedValue(strategy = GenerationType.IDENTITY)    //设置id生成方式为自增长
    private Integer id;
    @Column(name = "guarder_name")
    private String guarderName;
    @Column
    private String address;
    @Column(name = "phone_number")
    private String phoneNumber;
    @OneToOne(mappedBy = "contact")
    private Student student;

    public Contact() {
    }
```

```java
    public Contact(Integer id, String guarderName, String address, String phoneNumber, Student student) {
        this.id = id;
        this.guarderName = guarderName;
        this.address = address;
        this.phoneNumber = phoneNumber;
        this.student = student;
    }

    // get、set 方法

    @Override
    public String toString() {
        return "Contact{" +
                "id=" + id +
                ", guarderName='" + guarderName + '\'' +
                ", address='" + address + '\'' +
                ", phoneNumber='" + phoneNumber +
                '}';
    }
}
```

下面的示例展示一对多@OneToMany和多对一@ManyToOne的使用。

【示例4.2】 科目和教师的一对多关系

一个教师教授某一个科目,一个科目有多个教师。在教师表中创建关联字段,存储科目id来进行关联。

教师类Teacher.java代码如下:

```java
package com.example.school.pojo;

import javax.persistence.*;

@Entity
@Table(name = "teacher")
public class Teacher {
    @Id
    @GeneratedValue(strategy = GenerationType.IDENTITY)
    private Integer id;
    @Column
    private String name;
    @ManyToOne(cascade = CascadeType.PERSIST)
    @JoinColumn(name = "subject_id", referencedColumnName = "id")
    private Subject subject;

    public Teacher() {
    }

    public Teacher(Integer id, String name, Subject subject) {
        this.id = id;
        this.name = name;
        this.subject = subject;
    }
```

```java
        // get、set 方法
        public void setSubject(Subject subject) {
            this.subject = subject;
        }

        @Override
        public String toString() {
            return "Teacher{" +
                    "id=" + id +
                    ", name='" + name + '\'' +
                    ", subject=" + subject +
                    '}';
        }
}
```

科目类Subject.java代码如下：

```java
package com.example.school.pojo;

import javax.persistence.*;
import java.util.List;

@Entity
@Table(name = "subject")
public class Subject {
    @Id
    @GeneratedValue(strategy = GenerationType.IDENTITY)
    private Integer id;
    @Column
    private String name;
    @OneToMany(mappedBy = "subject", fetch = FetchType.EAGER)
    private List<Teacher> teachers;

    public Subject() {
    }

    public Subject(Integer id, String name, List<Teacher> teachers) {
        this.id = id;
        this.name = name;
        this.teachers = teachers;
    }

    // get、set 方法

    @Override
    public String toString() {
        return "Subject{" +
                "id=" + id +
                ", name='" + name + '\'' +
                '}';
    }
}
```

2. 使用 Spring Data JPA 提供的实现方法操作数据

对于基础的增删改查操作，Spring Data JPA已经为我们提供了接口和实现类，使用时只需要用自己的接口继承官方提供的接口，便可以完成增删改查的基本数据操作。

具体来讲，Spring Data JPA提供的接口中，最易于使用的接口有两个，分别是JpaRepository和JpaSpecificationExecutor，它们都是在org.springframework.data.jpa.repository包中定义的。

（1）通过接口JpaRepository可以实现对数据的增删改和基础的查询操作。通过接下来的3个示例演示接口JpaRepository的增删改操作。

【示例4.3】 使用JpaRepository保存Student数据

首先，创建接口StudentDao，继承JpaRepository，完整代码如下：

```
package com.example.school.dao;
import com.example.school.pojo.Student;
import org.springframework.data.jpa.repository.JpaRepository;
public interface StudentDao extends JpaRepository<Student,Integer> {
}
```

在测试类中执行方法，注入StudentDao，并编写调用JpaRepository.save(...)的方法，代码如下：

```
@Autowired
StudentDao studentDao;
@Test
/**
 * 增加一条 Student 信息及其关联的 Contact
 */
void saveStudentContact() {
    Student student = new Student();
    student.setName("马冬梅");
    student.setAge(16);
    Contact contact = new Contact();
    contact.setGuarderName("马冬");
    contact.setAddress("烽火台");
    student.setContact(contact);
    studentDao.save(student);
}
```

执行测试方法，数据库中的结果如图4.6和图4.7所示。

id	age	gender	name	contact_id
2	16	(Null)	马冬梅	2

id	address	guarder_name	phone_number
2	烽火台	马冬	(Null)

图 4.6 测试生成的 Student 数据　　图 4.7 测试生成的 Contact 数据

这里生成数据的ID为2，在下面的两个示例中会用到。

【示例4.4】 使用JpaRepository修改Student数据

进行更新前需要先进行查询,对查询出的数据修改后再次调用save方法,而save方法在ID存在时会先进行查询,若有数据则更新,具体代码如下:

```
@Test
/**
 * 修改 Student 信息
 */
void updateStudent() {
    Optional<Student> studentOpt = studentDao.findById(2);
    if(!studentOpt.isEmpty()){
        Student student = studentOpt.get();
        student.setAge(17);
        studentDao.save(student);
    }
}
```

【示例4.5】 使用JpaRepository删除Student数据

根据ID删除在前面插入的Student数据,具体代码如下:

```
@Test
/**
 * 根据 ID 删除 Student
 */
void deleteStudent(){
    studentDao.deleteById(2);
}
```

由于在实体类Student中定义的一对一关联关系注解OneByOne中,指定了级联操作cascade=CascadeType.PERSIST,只是插入数据时进行级联,所以这里只是删除了Student.ID为2的数据,不会删除Contact记录。

(2)另一个常用的接口JpaSpecificationExecutor是一个独立接口,因为其没有继承Repository,所以不能单独使用,只能结合其他Repository的继承接口使用,通常配合JpaRepository来使用。在接口JpaSpecificationExecutor中定义了5个方法、4个查询、1个计数方法。继承此接口时可以指定一个类型,这个类型用来指明查询结果的实体类。

接口JpaSpecificationExecutor的5个方法都需要Specification类型的参数,这个Specification类型的参数用来封装查询条件,其使用方式与Hibernate JPA非常相似。下面通过一个综合性的示例来体验Specification参数和接口JpaSpecificationExecutor的使用。

【示例4.6】 对Teacher数据分页查询并按ID排序

示例用到的数据库中的Teacher数据如图4.8所示,Subject数据如图4.9所示。

示例目标是查询教授科目为数学的老师,按ID由大到小排序,分页显示且每页5条,查询第1页。先创建用于处理Subject数据的接口SubjectDao,继承接口Specification和JpaSpecificationExecutor,并指定泛型。具体代码如下:

id	name	subject_id
1	王老师	1
2	李老师	2
3	赵老师	3
4	马老师	4
5	杜老师	5
6	钱老师	6
7	孙老师	3
8	周老师	3
9	吴老师	3
10	甄老师	3
11	宋老师	3

图 4.8　Teacher 示例数据

id	name
1	语文
2	英语
3	数学
4	物理
5	化学
6	生物

图 4.9　Subject 示例数据

```
package com.example.school.dao;

import com.example.school.pojo.Subject;
import org.springframework.data.jpa.repository.JpaRepository;
import org.springframework.data.jpa.repository.JpaSpecificationExecutor;

public interface SubjectDao extends JpaRepository<Subject, Integer>,
JpaSpecificationExecutor<Subject> {
}
```

然后，在测试类中注入SubjectDao对象，编写测试方法，创建参数以及分页和排序，具体代码如下：

```
@Autowired
TeacherDao teacherDao;
@Test
void findTeacherOfMath(){
    Sort sort = Sort.by(Sort.Direction.DESC,"id");
    Pageable pageable = PageRequest.of(0,5, sort);
    Specification<Teacher> specification = new Specification<Teacher>() {
        @Override
        public Predicate toPredicate(Root<Teacher> root, CriteriaQuery<?> query,
CriteriaBuilder criteriaBuilder) {
            return criteriaBuilder.equal(root.get("subject").get("name"),"数学
");
        }
    };
    Page<Teacher> teachers = teacherDao.findAll(specification,pageable);
    long totalElements = teachers.getTotalElements();
    System.out.println("总记录条数: " + totalElements);
    int totalPages = teachers.getTotalPages();
    System.out.println("总页数: " + totalPages);
    for (Teacher teacher : teachers) {
        System.out.println(teacher);
    }
}
```

上面的代码指定排序字段时，指定的是实体类的字段名。由于这里的id在表中和实体类中一致，因此在这里特意说明一下。

执行测试类，可以在控制台看到结果，如图4.10所示。

```
总记录条数：6
总页数：2
Teacher{id=11, name='宋老师', subject=Subject{id=3, name='数学'}}
Teacher{id=10, name='甄老师', subject=Subject{id=3, name='数学'}}
Teacher{id=9, name='吴老师', subject=Subject{id=3, name='数学'}}
Teacher{id=8, name='周老师', subject=Subject{id=3, name='数学'}}
Teacher{id=7, name='孙老师', subject=Subject{id=3, name='数学'}}
```

图4.10 打印分页信息

3．基于方法名称的查询

我们的接口继承JpaRepository，间接继承了Spring Data模块下的顶级接口Repository。继承Repository这个接口可以使用基于方法名称的命名规则查询。

在继承或间接继承接口Repository的接口中，按照一定的命名规则定义方法，框架会为其实现方法。方法名称分为3部分：

- findBy。
- 属性名称，要将首字母大写。
- 查询关键字，如Is、Equal、Like等，其中Is和Equal可以省略。

表4.3中列举了查询条件和示例。

表4.3 根据方法名查询命名关键字

查询关键字	方法名举例	说 明
Is	findByNameIs 的 Is 省略，为findByName	查询name字段为指定值，生成的SQL语句为：where name = ?
Equals	findByNameEquals的Equals省略，为findByName	查询name字段为指定值，生成的SQL语句为：where name = ?
Not	findByNameNot	查询name字段不为指定值的所有数据，生成的SQL语句为：Where name <> ?
Like	findByNameLike	like关键字查询，需要自己为参数拼接通配符，支持escape关键字的使用，生成的SQL语句为：where name like ?
StartingWith	findByNameStartingWith	使用like关键字查询，相当于在传入参数的结尾拼接上了通配符%，生成的SQL语句为：where name like ?
EndingWith	findByNameEndingWith	使用like关键字查询，相当于在传入参数的开头拼接上了通配符%，生成的SQL语句为：where name like ?
Containing	findByNameContaining	使用like关键字查询，相当于在传入参数的开头和结尾拼接上了通配符%，生成的SQL语句为：where name like ?
NotLike	findByNameNotLike	Like的否定形式，生成的SQL语句为：where name not like ?

（续表）

查询关键字	方法名举例	说明
IsNull	findByNameIsNull	查询name字段为null的数据，生成的SQL语句为：where name is null
IsNotNull	findByNameIsNotNull	查询name字段不为null的数据，生成的SQL语句为：where name is not null
NotNull	findByNameNotNull	查询name字段不为null的数据，生成的SQL语句为：where name is not null
Between	findByAgeBetween	需要传递两个参数，查询年龄在这两个参数（包含两个端点）之间的数据，生成的SQL语句为：where age between ? and ?
Or	findByNameIsOrAgeBetween	用于连接两个查询条件，查询name符合指定值或年龄符合范围的数据，生成的SQL语句为：where name=? or age between ? and ?
And	findByNameIsAndAgeBetween	用于连接两个查询条件，查询name符合指定值并且年龄符合范围的数据，生成的SQL语句为：where name=? and age between ? and ?
LessThan	findByAgeLessThan	查询age字段小于指定值的数据，生成的SQL语句为：where age<?
LessThanEqual	findByAgeLessThanEqual	查询age字段小于或等于指定值的数据，生成的SQL语句为：where age<=?
GreaterThan	findByAgeGreaterThan	查询age字段大于指定值的数据，生成的SQL语句为：where age>?
GreaterThanEqual	findByAgeGreaterThanEqual	查询age字段大于或等于指定值的数据，生成的SQL语句为：where age>=?
Before	findByAgeBefore	查询age字段小于指定值的数据，生成的SQL语句为：where age<?
After	findByAgeAfter	查询age字段大于指定值的数据，生成的SQL语句为：where age>?
OrderBy	findByOrderByIdDesc	将数据按id字段倒序排列，生成的SQL语句为：order by id desc
In	findByAgeIn	需要传入集合类型参数，查询数据在集合中的数据，生成的SQL语句为：where age in (?, ?)
NotIn	findByAgeNotIn	需要传入集合类型参数，查询数据不在集合中的数据，生成的SQL语句为：where age not in (?, ?)
IgnoreCase	findByNameIsIgnoreCase	忽略传入参数的大小写，SQL语句中会将字段和参数都转为大写判断，生成的SQL语句为：where upper(name)=upper(?)

这种查询方式简单易用，但是缺点是在条件多和条件复杂时，这种方法名太长，易读性差。下面是一个查询示例。

【示例4.7】 查询name为"小花"且年龄在13~17的Student数据

先创建接口StudentQueryDao,继承接口Repository,并根据查询条件创建方法,代码如下:

```java
package com.example.school.dao;

import com.example.school.pojo.Student;
import org.springframework.data.repository.Repository;

import java.util.List;

public interface StudentQueryDao extends Repository<Student, Integer> {
    List<Student> findByNameIsAndAgeBetween(String name,int less, int great);
}
```

在测试类中注入接口StudentQueryDao的对象,然后调用此方法,代码如下:

```java
@Autowired
StudentQueryDao studentQueryDao;
@Test
void testQueryDao(){
    List<Student> students = studentQueryDao.findByNameIsAndAgeBetween("小花",13,17);
    for (Student student : students) {
        System.out.println(student);
    }
}
```

4. 使用@Query 注解

@Query注解所在的包为org.springframework.data.jpa.repository,可以使用JPQL语句和SQL语句。注解@Query中定义的元数据如下:

- value:指定用于查询的语句。
- countQuery:使用分页时,指定用于计数的语句。
- countProjection:使用分页以及 Projection 时,指定用于计数的语句。
- nativeQuery:标注绑定的查询语句是否为原生 SQL 语句,默认值为 false,也就是默认绑定的查询语句为 JPQL 格式。
- name:使用分页时,指定用于查询的 named sql 的名称。
- countName:使用分页时,指定用于计数的 named sql 的名称。

指定查询语句使用value属性,如果使用name属性指定named sql,在找不到named sql时,框架会把方法当作基于方法名查询的命名规则来解析方法名。最后,我们通过下面的分页查询示例来进一步认识注解@Query的使用。

【示例4.8】 使用@Query根据科目名称查询教师信息

接口TeacherDao继承接口Repository或者Repository的子接口,这里继承JpaRepository。在TeacherDao中定义方法,方法名不受命名规则限制,具体代码如下:

```java
package com.example.school.dao;
```

```
import com.example.school.pojo.Teacher;
import org.springframework.data.domain.Page;
import org.springframework.data.domain.Pageable;
import org.springframework.data.jpa.repository.JpaRepository;
import org.springframework.data.jpa.repository.JpaSpecificationExecutor;
import org.springframework.data.jpa.repository.Query;
public interface TeacherDao extends JpaRepository<Teacher, Integer>,
JpaSpecificationExecutor<Teacher> {
    /**
     * 根据科目名称查询老师
     * @param subjectName 科目名称
     * @return
     */
    @Query(
        value = "select * from teacher t inner join subject s on t.subject_id = s.id where s.name = ?1 ",
        countQuery = "select count(*) from teacher t inner join Subject s on t.subject_id = s.id where s.name = ?1 ",
        nativeQuery = true
    )
    Page<Teacher> findTeacherOfSubject(String subjectName, Pageable pageable);
}
```

测试代码如下：

```
@Test
void findTeacherOfSubject() {
    Sort sort = Sort.by(Sort.Direction.DESC, "id");
    Pageable pageable = PageRequest.of(0, 5, sort);
    Page<Teacher> teacherPage = teacherDao.findTeacherOfSubject("数学", pageable);
    if (!teacherPage.isEmpty()) {
        for (Teacher teacher : teacherPage) {
            System.out.println(teacher);
        }
    }
}
```

另外，通过@Query还可以完成数据更新。进行更新时，需要在方法上添加@Modifying，以及需要@Transactional提供事务。注解@Transactional可以放到接口方法上，也可以放到测试方法上，放到测试方法上默认会回滚。如果想要不回滚，使用注解@Rollback(false)修饰测试方法即可。

【示例4.9】 使用@Query修改教师姓名

在接口TeacherDao中创建用于修改教师姓名的方法，并使用@Query注解指定要执行的JPQL语句，代码如下：

```
@Query(value = "update Teacher set name = ?2 where id = ?1")
@Modifying
```

```
@Transactional
int modifyTeacher(int id, String newName);
```

测试代码如下：

```
@Test
// @Rollback(false)  由于@Transactional标记在接口方法上，因此这里事务会提交
void testModify(){
    teacherDao.modifyTeacher(1,"王帅");
}
```

5．自定义 Repository 接口

自定义的Repository接口和Spring Data JPA的org.springframework.data.repository.Repository.Repository接口是同级别的。需要自己实现接口，通过实现类的命名规则，框架可以将我们的实现类对象一并代理，注入给已存在的Dao接口，使用起来非常方便。

接口实现类命名为接口名字拼接上Impl。在接口实现类中，使用EntityManager()来操作数据，这和JpaRepository、JpaSpecificationExecutor的实现类SimpleJpaRepository以及SimpleJpaSpecificationExecutor是一样的。接下来，通过下面的示例来进一步认识自定义Repository接口。

【示例4.10】　自定义TeacherRepository

首先，创建一个repository包用于存放自定义接口和实现类。创建一个接口TeacherRepository放到新创建的repository包中，并在接口中定义根据id查询教师的方法，代码如下：

```
package com.example.school.repository;
import com.example.school.pojo.Teacher;
public interface TeacherRepository{
    /**
     * 根据id查询教师
     * @param id 要查询的教师
     * @return
     */
    Teacher queryByTeacherId(Integer id);
}
```

然后，根据命名规则创建实现类TeacherRepositoryImpl，从容器中注入EntityManager对象，并实现接口TeacherRepository，代码如下：

```
package com.example.school.repository;
import com.example.school.pojo.Teacher;
import javax.persistence.EntityManager;
import javax.persistence.PersistenceContext;
public class TeacherRepositoryImpl implements TeacherRepository{
    @PersistenceContext
    EntityManager entityManager;
```

```
        @Override
        public Teacher queryByTeacherId(Integer id) {
            return entityManager.find(Teacher.class,id);
        }
    }
```

最后，让接口TeacherDao继承我们自定义的Repository接口TeacherRepository，代码如下：

```
public interface TeacherDao extends JpaRepository<Teacher, Integer>,
JpaSpecificationExecutor<Teacher>, TeacherRepository {
```

无须修改其他代码，在测试类中可以通过TeacherDao的对象使用TeacherRepository的方法，测试代码如下：

```
@Test
void testCustomRepository(){
    // 这里查询了 ID 为 10 的 Teacher 信息
    Teacher teacher = teacherDao.queryByTeacherId(10);
    System.out.println(teacher);
}
```

4.4 整合MyBatis框架

MyBatis是目前最为流行的持久层框架之一，使用MyBatis几乎可以避免所有的JDBC代码和手动设置参数。本节将介绍如何在Spring Boot中整合MyBatis框架和相关的自动配置。

4.4.1 MyBatis 简介

MyBatis是一款流行的持久层框架，支持自定义SQL、存储过程和高级映射。通过使用MyBatis，几乎可以省略所有JDBC代码、设置参数以及封装结果集的工作。MyBatis可以通过简单的XML或注解来配置和映射原始类型、接口和Java POJO（Plain Old Java Objects，普通老式Java对象）为数据库中的记录。MyBatis的Logo如图4.11所示。

图 4.11　MyBatis 官方 Logo

4.4.2 MyBatis 的配置

要使用MyBatis，需要将mybatis-x.x.x.jar文件置于类路径（classpath）下。由于MyBatis提供了与Spring Boot整合的Starter，因此直接引入MyBatis提供的Starter即可。

（1）将下面的依赖代码添加到项目的pom.xml文件中：

```xml
<dependency>
    <groupId>org.mybatis.spring.boot</groupId>
    <artifactId>mybatis-spring-boot-starter</artifactId>
    <version>2.1.4</version>
</dependency>
```

(2)使用Maven下载相关依赖,等待下载完成后,我们可以在项目External Libraries中看到MyBatis相关的JAR包,如图4.12所示。

图4.12 MyBatis 相关 JAR 包

(3)接下来创建MyBatis用到的POJO实体类、Mapper接口和Mapper.xml文件,文件目录结构如图4.13所示。

图4.13 MyBatis 相关代码目录结构

User.java文件内容如下:

```java
package com.example.demo.pojo;
public class User {
    private int id;
    private String name;
    private int age;
    private String address;
    public User() {
    }
    public User(int id, String name, int age, String address) {
        this.id = id;
        this.name = name;
        this.age = age;
        this.address = address;
    }
    public int getId() {
        return id;
    }
    public void setId(int id) {
        this.id = id;
    }
    public String getName() {
        return name;
    }
    public void setName(String name) {
        this.name = name;
    }
    public int getAge() {
        return age;
```

```
        }
        public void setAge(int age) {
            this.age = age;
        }
        public String getAddress() {
            return address;
        }
        public void setAddress(String address) {
            this.address = address;
        }
        @Override
        public String toString() {
            return "User{" +
                    "id=" + id +
                    ", name='" + name + '\'' +
                    ", age=" + age +
                    ", address='" + address + '\'' +
                    '}';
        }
    }
```

UserMapper.java代码如下:

```
package com.example.demo.dao;
import com.example.demo.pojo.User;
import org.apache.ibatis.annotations.Mapper;
import org.springframework.stereotype.Repository;
import java.util.List;

@Repository
@Mapper
public interface UserMapper {
    int add(User user);
    int deleteById(int id);
    int updateById(User user);
    List<User> queryAll();
}
```

Mapper.xml文件内容如下:

```
<?xml version="1.0" encoding="UTF-8"?>
<!DOCTYPE mapper PUBLIC "-//mybatis.org//DTD Mapper 3.0//EN"
"http://mybatis.org/dtd/mybatis-3-mapper.dtd">
    <mapper namespace="com.example.demo.dao.UserMapper">
        <insert id="add" parameterType="com.example.demo.pojo.User">
            insert into user(id,name,age,address) values (#{id},#{name},#{age},#{address})
        </insert>
        <delete id="deleteById" parameterType="int">
            delete from user where id=#{id}
        </delete>
        <update id="updateById" parameterType="com.example.demo.pojo.User">
```

```
            update user set name=#{name},age=#{age},address=#{address} where id=#{id}
        </update>
        <select id="queryAll" resultType="com.example.demo.pojo.User">
            select * from user
        </select>
</mapper>
```

（4）至此，完成了使用MyBatis对MySQL中user表增删改查操作的代码。user表使用4.2节中创建的表。

4.4.3 Spring Boot 整合 MyBatis

MyBatis官方提供了与Spring Boot整合的Starter，在4.4.2节中已引入pom.xml中，这里无须重复引入。此时只需要在application.yaml中添加配置，代码如下：

```
mybatis:
  mapper-locations: "classpath:mapper/*.xml"
```

此配置用于指定MyBatis扫描mapper XML路径。

mapper XML的namespace显示绑定接口com.example.demo.dao.UserMapper，并且语句id需要与@Mapper的接口中的方法名一致，且参数一致。

至此，已完成Spring Boot与MyBatis整合的所有工作。接下来我们进行测试，观察执行结果。

（1）首先在DemoApplicationTests.java中添加代码，添加4个方法，分别测试我们在UserMapper接口中定义的方法。

```
@Autowired
UserMapper userMapper;
@Test
void testMybatisAdd(){
    int addCount = userMapper.add(new User(3, "白鸽", 26, "隆平路"));
    System.out.println("增加用户个数："+addCount);
}
@Test
void testMybatisQueryAll(){
    List<User> users = userMapper.queryAll();
    for(User user : users){
        System.out.println("用户："+user);
    }
}
@Test
void testMybatisUpdate(){
    int updateCount = userMapper.updateById(new User(3, "白鸽", 22, "昭庆街"));
    System.out.println("修改用户个数："+updateCount);
}
@Test
```

```
void testMybatisDelete(){
    int deleteCount = userMapper.deleteById(3);
    System.out.println("删除用户个数："+deleteCount);
}
```

（2）执行之前的数据库记录，如图4.14所示。

（3）执行方法testMybatisAdd()，控制台输出：

```
增加用户个数：1
```

数据库记录增加一条，信息如图4.15所示。

id	name	age	address
1	张三	18	新华街
2	李四	22	康庄路

图 4.14　测试程序执行前的记录

id	name	age	address
1	张三	18	新华街
2	李四	22	康庄路
3	白鸽	26	隆平路

图 4.15　测试程序 testMybatisAdd 执行后的记录

（4）再执行方法testMybatisQueryAll()，控制台打印出3条用户信息，与数据库记录一致。

```
用户：User{id=1, name='张三', age=18, address='新华街'}
用户：User{id=2, name='李四', age=22, address='康庄路'}
用户：User{id=3, name='白鸽', age=26, address='隆平路'}
```

（5）然后执行方法testMybatisUpdate()，控制台打印：

```
修改用户个数：1
```

数据库中的信息仅id为3的白鸽的age和address发生变化，另外两条记录无变化，如图4.16所示。

（6）最后执行testMybatisDelete()，删除id为3的白鸽，控制台输出：

```
删除用户个数：1
```

查看数据库，只剩下两条记录，如图4.17所示。

id	name	age	address
1	张三	18	新华街
2	李四	22	康庄路
3	白鸽	22	昭庆街

图 4.16　测试程序 testMybatisUpdate 执行后的记录

id	name	age	address
1	张三	18	新华街
2	李四	22	康庄路

图 4.17　测试程序 testMybatisDelete 执行后的记录

4.4.4　MyBatis 的其他配置

在上一节对Spring Boot与MyBatis整合时，使用到了配置属性mapper-locations，用于指定mapper xml文件的路径。这一节介绍MyBatis定义的其他属性。

先介绍如何查看MyBatis定义了哪些属性。在External Libraries中找到mybatis-spring-boot-autoconfigre，在其中找到MybatisProperties类，如图4.18所示。

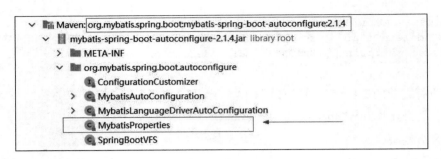

图 4.18　MybatisProperties 类路径

在这个类中定义的字段可以在application.yaml中配置，其中String[] mapperLocations在上一节已经用过，它的类型是String[]，所以可以配置多个mapper文件路径。

- String configLocation：用于指定 MyBatis 的 XML 配置文件。在已有 MyBatis 项目中引入 Spring Boot 时使用最为方便，不用修改已有的 MyBatis 配置文件，使用该参数指定即可生效。
- String typeAliasesPackage：用于指定搜索类型别名的包名。一般对应我们的实体类所在的包，会自动取对应包中不包括包名的简单类名作为类型的别名。指定多个包名时，使用分隔符分隔，分隔符有,、;、\t、\n。
- Class<?> typeAliasesSuperType：配合 typeAliasesPackage 使用，指定仅扫描以此类为父类的类型。
- String typeHandlersPackage：用于指定需要将类型注册为对应的 TypeHandler 的包名。指定多个包名时，使用分隔符分隔，分隔符有,、;、\t、\n。
- boolean checkConfigLocation：指定检查 MyBatis 的 XML 文件是否存在，默认值为 false。
- ExecutorType executorType：指定 org.mybatis.spring.SqlSessionTemplate 的执行模式，有 SIMPLE、REUSE 和 BATCH 可选。
- Class<? extends LanguageDriver> defaultScriptingLanguageDriver：指定默认的脚本语言驱动器，在 mybatis-spring 2.0.2 及以后的版本可以使用。
- Properties configurationProperties：指定外部化 MyBatis Properties 配置，通过该配置可以抽离配置，实现不同环境的配置。
- Configuration configuration：通过 Configuration 对象来自定义默认配置。如果使用了 String configLocation 属性，那么这个属性不会被用到。

4.5　实战——商品信息管理小系统

本节将模拟开发一个小商店后台商品管理系统，实现增加商品、删除商品、商品上架、商品下架和修改商品的信息等功能。

第 4 章 使用 Spring Boot 进行数据库开发

（1）分析需求，定义表结构

首先，有一张表，用于存储商品信息。商品信息有商品ID、商品名称、供货商、采购单价、销售单价、上架状态和库存数量。其中，供货商字段我们使用外键关联到供货商表。供货商表存储供货商信息，有供货商ID、供货商名称和供货商地址。

（2）技术选型

根据需求分析，共有两张表，用于实现商品到供应商的多对一关联关系，以及相对简单的增删改查操作。数据库选择MySQL存储，持久层框架这里不妨选择Spring Data JPA，可以避免写SQL语句，以简化开发。

页面使用动态HTML技术，使用第3章介绍的Thymeleaf模板引擎实现。

（3）创建项目

使用Spring Initializr来创建项目，项目元信息如图4.19所示。

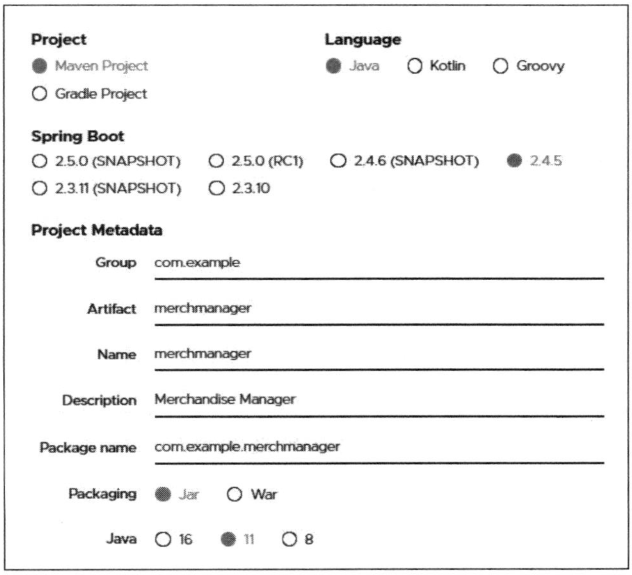

图 4.19　项目元信息

根据技术选型，项目依赖信息如图4.20所示。

创建完成后，还是使用IDEA导入项目，导入后删除application.properties，创建配置文件application.yaml，目录结构如图4.21所示。

（4）配置数据源信息

在application.yaml中添加datasource的配置，作者的配置如下：

```
spring:
  datasource:
    url: "jdbc:mysql://127.0.0.1:3306/db"
    username: "dbuser"
    password: "123456"
```

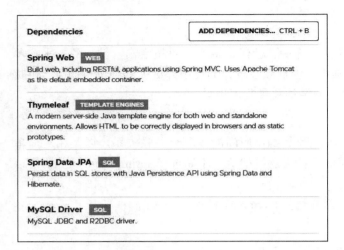

图 4.20　项目依赖信息

图 4.21　项目目录结构

运行MerchmanagerApplication的main方法，如果配置正确，可以看到程序启动成功的控制台日志输出，如图4.22所示。

图 4.22　项目启动成功

（5）创建商品和供应商两个实体类，并创建对应的映射关系

在程序主包下创建pojo包，然后在pojo包下创建实体类Merchandise和Supplier，分别对应商品和供应商，如图4.23所示。

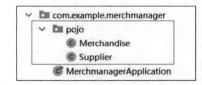

根据上面（1）中的分析，为两个实体类添加字段，并创建关联关系。

图 4.23　创建实体类

Merchandise.java的代码具体如下：

```java
package com.example.merchmanager.pojo;

import javax.persistence.*;

@Entity
@Table(name = "merchandise")
public class Merchandise {
    @Id
    @GeneratedValue(strategy = GenerationType.IDENTITY)
    private Integer id;
    private String name;
    @ManyToOne(cascade = CascadeType.PERSIST, fetch = FetchType.EAGER)
    @JoinColumn(name = "supplier_id")
    private Supplier supplier;
```

```java
        @Column(name = "purchase_price")
        private Double purchasePrice;
        private Double price;
        private String onSaleFlag;
        private Integer stock;

        public Merchandise() {
        }

        public Merchandise(Integer id, String name, Supplier supplier, Double purchasePrice, Double price, String onSaleFlag, Integer stock) {
            this.id = id;
            this.name = name;
            this.supplier = supplier;
            this.purchasePrice = purchasePrice;
            this.price = price;
            this.onSaleFlag = onSaleFlag;
            this.stock = stock;
        }

        // get、set方法

        @Override
        public String toString() {
            return "Merchandise{" +
                "id=" + id +
                ", name='" + name + '\'' +
                ", supplier=" + supplier +
                ", purchasePrice=" + purchasePrice +
                ", price=" + price +
                ", onSaleFlag=" + onSaleFlag +
                ", stock=" + stock +
                '}';
        }
    }
```

Supplier.java的代码具体如下：

```java
package com.example.merchmanager.pojo;

import javax.persistence.*;
import java.util.List;

@Entity
@Table(name = "supplier")
public class Supplier {
    @Id
    @GeneratedValue(strategy = GenerationType.IDENTITY)
    private Integer id;
    private String name;
    private String address;

    // get、set方法

    @Override
```

```
    public String toString() {
        return "Supplier{" +
                "id=" + id +
                ", name='" + name + '\'' +
                ", address='" + address + '\'' +
                '}';
    }
}
```

（6）创建持久层接口

创建dao包，在其中创建接口MerchandiseDao和SupplierDao，各自继承Spring Data JPA接口JpaRepository和JpaSpecificationExecutor。

MerchandiseDao.java接口代码如下：

```
package com.example.merchmanager.dao;

import com.example.merchmanager.pojo.Merchandise;
import org.springframework.data.jpa.repository.JpaRepository;
import org.springframework.data.jpa.repository.JpaSpecificationExecutor;

public interface MerchandiseDao extends JpaRepository<Merchandise, Integer>, JpaSpecificationExecutor<Merchandise> {
}
```

SupplierDao.java接口代码如下：

```
package com.example.merchmanager.dao;

import com.example.merchmanager.pojo.Merchandise;
import com.example.merchmanager.pojo.Supplier;
import org.springframework.data.jpa.repository.JpaRepository;
import org.springframework.data.jpa.repository.JpaSpecificationExecutor;

public interface SupplierDao extends JpaRepository<Supplier, Integer>, JpaSpecificationExecutor<Supplier> {
}
```

（7）分析需要创建的页面

- 主页面：显示"商品管理"和"供应商管理"两个链接。单击"商品管理"链接进入商品管理页面，单击"供应商管理"链接进入供应商管理页面。
- 商品管理页面：有商品列表以及"增加""删除"和"编辑"3个按钮。
- 供应商管理页面：有供应商列表以及"增加""删除"和"编辑"3个按钮。

（8）分析页面用到的接口

主页命名为index.html，纯粹是HTML文件，不需要模板引擎处理，放到/static目录下。

商品管理页面需要有列表接口，接口定义为/merchandise。接口返回列表页面，页面命名为merchandise.html。此外，还需要商品信息的增加、编辑和删除3个接口。"增加"和"编辑"共用一个接口/merchandise/save，"删除"接口为/merchandise/delete。

供应商管理页面主页为列表页，接口定义为/supplier，接口返回内容为supplier.html。另外，将供应商信息的"增加"和"编辑"共用接口定义为/supplier/save,"删除"接口定义为/supplier/delete。

(9) 编写 HTML 页面

系统主页index.html代码如下:

```
<!DOCTYPE html>
<html lang="en">
<head>
    <meta charset="UTF-8">
    <title>商品信息管理系统-主页</title>
</head>
<body>
<h1>商品信息管理系统</h1>
<ul>
    <li>
        <a href="/merchandise">商品管理</a>
    </li>
    <li>
        <a href="/supplier">供应商管理</a>
    </li>
</ul>
</body>
</html>
```

在开发商品管理页面和供应商管理页面之前,引入JS库jQuery文件和自定义的一个CSS文件。在static目录下创建js和css目录,分别放入jquery-3.6.0.min.js和total.css文件。

jquery-3.6.0.min.js文件可以从jQuery官网下载到。total.css文件内容如下:

```
th{
    border:1px solid #ffe812;
    vertical-align: middle;
    text-align: center
}
table td{
    border:1px solid #ffe812;
    vertical-align: middle;
    text-align: center
}
#smallWindow {
    display: none;
    position: fixed;
    left: 0;
    top: 0;
    width: 100%;
    height: 100%;
    background-color: rgba(0,0,0,0.5);
}
#smallWindowContent {
    background:#eeeeee;
```

```css
    width: 22%;
    z-index: 100;
    margin: 12% auto;
    overflow: auto;
}
span {
    color: white;
    padding-top: 12px;
    cursor: pointer;
    padding-right: 15px;
}
#contentdiv {
    background:#eeeeee;
    margin: auto;
    height: 280px;
    padding: 0 20px;
}
#closediv {
    padding: 1px;
    background: #afbba9;
}
#closebtn {
    float: right;
    font-size: 40px;
}
```

下面开发商品管理页面，merchandise.html部分代码如下：

```html
<!DOCTYPE html>
<html lang="en" xmlns:th="https://www.thymeleaf.org">
<head>
    <meta charset="UTF-8">
    <title>商品信息管理系统-商品管理</title>
    <script src="/js/jquery-3.6.0.min.js"></script>
    <link rel="stylesheet" type="text/css" href="/css/total.css" />
<script>
    var is_edit = false;
    var edit_tr_id ;
    function add(){
        $("#merch_id").val("");
        $("#merch_name").val("");
        $("#merch_supplier_id").val("");
        $("#merch_purchase_price").val("");
        $("#merch_price").val("");
        $("#merch_on_sale_flag").val("");
        $("#merch_stock").val("");
        is_edit = false;
        showSmallWindow("新增");
        return;
    }
    ...
```

```html
<!-- 以下为弹窗部分 -->
<div id="smallWindow">
    <div id="smallWindowContent">
        <div id="closediv">
            <span id="closebtn" onclick="closeSmallWindow()">×</span>
            <h4 id="smallWindowTitle">弹窗头部</h4>
        </div>
        <div id="contentdiv">
            <table>
                <tr id="idtr">
                    <td>ID</td>
                    <td><input id="merch_id" type="text" disabled></td>
                </tr>
                <tr>
                    <td>商品名称</td>
                    <td><input id="merch_name" type="text"></td>
                </tr>
                <tr id="merch_supplier_id_tr">
                    <td>供应商ID</td>
                    <td><input id="merch_supplier_id" type="text"></td>
                </tr>
                <tr>
                    <td>采购单价</td>
                    <td><input id="merch_purchase_price" type="text"></td>
                </tr>
                <tr>
                    <td>销售单价</td>
                    <td><input id="merch_price" type="text"></td>
                </tr>
                <tr>
                    <td>销售状态</td>
                    <td><input id="merch_on_sale_flag" type="text"></td>
                </tr>
                <tr>
                    <td>商品库存</td>
                    <td><input id="merch_stock" type="text"></td>
                </tr>
                <tr>
                    <td colspan="2">
                        <button onclick="save(this)">保存</button>
                    </td>
                </tr>
            </table>
        </div>
    </div>
</div>
</body>
</html>
```

开发供应商管理页面，supplier.html部分代码如下：

```html
<!DOCTYPE html>
<html lang="en" xmlns:th="https://www.thymeleaf.org">
<head>
    <meta charset="UTF-8">
    <title>商品信息管理系统-供应商管理</title>
    <script src="/js/jquery-3.6.0.min.js"></script>
    <link rel="stylesheet" type="text/css" href="/css/total.css" />
    <style>
        #contentdiv {
            background:#eeeeee;
            margin: auto;
            height: 200px;
            padding: 0 20px;
        }
    </style>

    <script>
    var is_edit = false;
    var edit_tr_id ;
    function add(){
        $("#supplier_id").val("");
        $("#supplier_name").val("");
        $("#supplier_address").val("");
        is_edit = false;
        showSmallWindow("新增");
        return;
    }
    function add_callBack(data){
        var tBody = $("#list");
        tBody.append("<tr><td>"+data.id+"</td><td>"+data.name+"</td><td>"+data.address+"</td><td>"
            +"<button onclick='edit(this)'>编辑</button> "
            +"<button onclick='deletee(this)' th:supplier_id='${supplier.id}' >删除</button>"
            +"</td></tr>");
    }
    function edit_callBack(data){
        var td = edit_tr_id;
        td.text(data.id);
        td = td.next();
        td.text(data.name);
        td = td.next();
        td.text(data.address);
    }
</html>
```

（10）开发接口

在第（6）步中已经创建好了持久层接口，在这里只需要创建Controller层，对持久层接口进行调用即可完成项目。

创建controller包，在controller包下创建两个Controller类，用于开发商品信息管理接口和供应商信息管理接口。

在controller包下创建商品信息管理控制类MerchandiseController.java，代码如下：

```java
package com.example.merchmanager.controller;

import com.example.merchmanager.dao.MerchandiseDao;
import com.example.merchmanager.dao.SupplierDao;
import com.example.merchmanager.pojo.Merchandise;
import org.springframework.beans.factory.annotation.Autowired;
import org.springframework.stereotype.Controller;
import org.springframework.web.bind.annotation.RequestMapping;
import org.springframework.web.bind.annotation.RequestParam;
import org.springframework.web.bind.annotation.ResponseBody;

import java.util.List;
import java.util.Map;

@Controller
@RequestMapping("/merchandise")
public class MerchandiseController {
    @Autowired
    MerchandiseDao merchandiseDao;
    @Autowired
    SupplierDao supplierDao;
    @RequestMapping
    public String listPage(Map<String, List<Merchandise>> map){
        List<Merchandise> merchandises = merchandiseDao.findAll();
        map.put("merchandises", merchandises);
        return "merchandise";
    }
    /**
     * 没有id时新增，有id时进行更新操作
     */
    @RequestMapping("save")
    @ResponseBody
    public Merchandise save(Integer id, String name,
                        @RequestParam(name = "supplier_id",required = false) Integer supplierId,
                        @RequestParam(name = "purchase_price",required = false) Double purchasePrice,
                        Double price,
                        @RequestParam(name = "on_sale_flag",required = false) String onSaleFlag,
                        Integer stock){
        Merchandise merchandise = new Merchandise();
        merchandise.setId(id);
        merchandise.setName(name);
        merchandise.setSupplier(supplierDao.findById(supplierId).get());
        merchandise.setPurchasePrice(purchasePrice);
        merchandise.setPrice(price);
```

```java
        merchandise.setOnSaleFlag(onSaleFlag);
        merchandise.setStock(stock);
        return merchandiseDao.save(merchandise);
    }
    @RequestMapping("delete")
    @ResponseBody
    public String delete(Integer id){
        merchandiseDao.deleteById(id);
        return "success";
    }
}
```

创建供应商管理类SupplierController.java,代码如下:

```java
package com.example.merchmanager.controller;

import com.example.merchmanager.dao.SupplierDao;
import com.example.merchmanager.pojo.Supplier;
import org.springframework.beans.factory.annotation.Autowired;
import org.springframework.stereotype.Controller;
import org.springframework.web.bind.annotation.RequestMapping;
import org.springframework.web.bind.annotation.ResponseBody;

import java.util.List;
import java.util.Map;

@Controller
@RequestMapping("/supplier")
public class SupplierController {
    @Autowired
    SupplierDao supplierDao;
    @RequestMapping
    public String listPage(Map<String, List<Supplier>> map){
        List<Supplier> suppliers = supplierDao.findAll();
        map.put("suppliers", suppliers);
        return "supplier";
    }
    /**
     * 没有id时新增,有id时进行更新操作
     */
    @RequestMapping("save")
    @ResponseBody
    public Supplier save(Integer id, String name, String address){
        Supplier supplier = new Supplier();
        supplier.setId(id);
        supplier.setName(name);
        supplier.setAddress(address);
        return supplierDao.save(supplier);
    }
    @RequestMapping("delete")
    @ResponseBody
    public String delete(String id){
```

```
            supplierDao.deleteById(Integer.parseInt(id));
            return "success";
    }
}
```

（11）运行项目，观察效果

启动项目，在浏览器中访问系统主页http://127.0.0.1:8080/index.html，效果如图4.24所示。单击"供应商管理"链接，进入供应商管理页面，如图4.25所示。

图4.24　系统主页

图4.25　供应商管理页面

单击"添加供应商"按钮，弹出供应商信息录入窗口，录入信息，单击"保存"按钮，如图4.26所示。

图4.26　新增供应商

单击"确定"按钮，录入窗口自动隐藏，并且列表显示新录入的数据，如图4.27所示。

图4.27　添加供应商后列表页显示

单击浏览器中的返回按钮，返回系统主页。再单击"商品管理"链接进入商品管理页面，如图4.28所示。

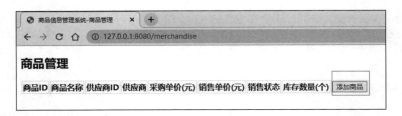

图 4.28 商品管理页面

单击"添加商品"按钮,显示商品新增窗口,录入商品信息。这里供应商ID填写"1",这是刚刚添加的供应商。单击"保存"按钮,如图4.29所示。

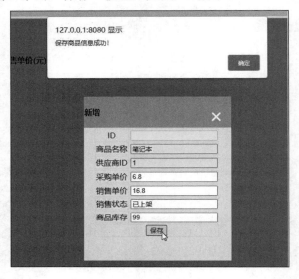

图 4.29 创建商品

单击"确定"按钮,录入信息窗口被隐藏,商品列表中显示新增加的商品信息,如图4.30所示。

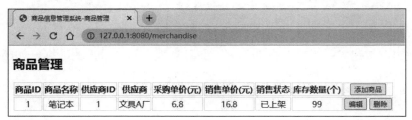

图 4.30 添加成功后列表显示商品信息

另外,单击"编辑"按钮会显示商品信息编辑窗口;单击"删除"按钮会删除所在行的商品。这就是整个示例项目的运行效果。

第 5 章

Spring Boot与Redis

Redis是先进的key-value数据库，它和Memcached类似，同时也支持更丰富的数据类型。由于其高性能的特色，Redis常常成为项目中做缓存的不二之选。本章将介绍如何在Spring Boot项目中方便地操作Redis数据库。

本章主要涉及的知识点有：

- ⌘ 使用spring-data-redis操作Redis数据库。
- ⌘ 如何在项目Spring Boot项目中使用Redis。
- ⌘ Redis的一些特殊用法。
- ⌘ Spring缓存注解的使用。

5.1 使用spring-data-redis操作Redis

项目spring-data-redis与Redis的关系就像是项目spring-data-jdbc与关系型数据库的关系一样，项目spring-data-redis是对Redis官方提供的API的进一步封装。本书中介绍的在Spring Boot项目中使用Redis，也主要是通过spring-data-redis的使用来操作Redis，因此本节将介绍spring-data-redis项目。

5.1.1 Spring Data Redis 项目的设计

通过消除冗余工作和样板代码，Spring Data Redis框架使得在Spring应用程序中编写使用Redis键值存储变得容易。Spring Data Redis提供了在Spring应用程序配置和访问Redis的简单方式。

对于和Redis的交互，Spring Data Redis提供了低水平和高水平的抽象，来简化开发者对基础性操作的关注。大部分需求都可以通过高水平的抽象来实现，官方也推荐优先使用高水平抽象的API。同时，在任何使用高水平抽象的位置也都可以使用低水平抽象，甚至Redis原生API。

1. RedisConnection 和 RedisConnectionFactory 接口

Spring Data Redis在org.springframework.data.redis.connection包下提供了接口RedisConnection和RedisConnectionFactory用于创建和管理与Redis服务的连接。

RedisConnection接口提供了连接Redis的核心模块，以及将底层库的异常包装为Spring统一的DAO异常。使用包装后的异常在切换连接器时不再需要修改代码。

Spring Data Redis支持的连接器有Lettuce和Jedis。下面分别介绍对于Lettuce和Jedis的配置，配置过程中在未使用Spring Boot时需要的步骤，仅做了解即可；在使用Spring Boot提供的Starter后，配置会更简单。

2. 配置 Lettuce 连接器

Lettuce是开源的基于Netty的Redis连接器，也是Spring Boot 2.0以后默认的Redis连接器。使用Lettuce需要先在项目中引入对于Lettuce的依赖，配置代码如下：

```xml
<dependencies>
  <dependency>
    <groupId>io.lettuce</groupId>
    <artifactId>lettuce-core</artifactId>
    <version>6.1.1.RELEASE</version>
  </dependency>
</dependencies>
```

使用代码配置方式为容器中注入Lettuce的Redis连接工厂，代码如下：

```java
@Configuration
class AppConfig {
    @Bean
    public LettuceConnectionFactory redisConnectionFactory() {
        return new LettuceConnectionFactory(new RedisStandaloneConfiguration("server", 6379));
    }
}
```

【示例5.1】 在Redis中保存键值对

首先按上面的配置引入Lettuce依赖，然后在代码中配置LettuceConnectionFactory。这里作者使用的是在Ubuntu子系统中安装的单节点Redis，IP地址为本机，具体代码如下：

```java
import org.springframework.context.annotation.Bean;
import org.springframework.context.annotation.Configuration;
import org.springframework.data.redis.connection.RedisStandaloneConfiguration;
import org.springframework.data.redis.connection.lettuce.LettuceConnectionFactory;

@Configuration
public class Config {
    @Bean
    public LettuceConnectionFactory redisConnectionFactory() {
```

```
        return new LettuceConnectionFactory(new RedisStandaloneConfiguration
("localhost", 6379));
    }
}
```

然后，使用测试类在Redis中保存键值对数据，具体代码如下：

```
@Autowired
RedisConnectionFactory redisConnectionFactory;
@Test
void testConnection(){
    RedisConnection connection = redisConnectionFactory.getConnection();
    connection.set("zhangsan".getBytes(),"河北".getBytes());
    connection.close();
}
```

最后，通过redis-cli查看Redis中保存的数据。为了使中文正常显示，在启动redis-cli时使用参数"--raw"，命令如下：

```
redis-cli --raw
```

通过命令"get zhangsan"获取存储的数据，如图5.1所示。至此，可以看到数据已经被存储到了Redis中。

图 5.1　Redis get 命令

上面使用的lettuce-core会被spring-boot-starter-data-redis来管理，所以在使用Spring Boot整合Redis时无须显式引入lettuce-core。

3．配置 Jedis 连接器

Jedis是Java连接Redis服务时Redis官方推荐的连接器，是由社区提供的开源连接器。相比于Lettuce，性能低，而且非线程安全。下面的配置也仅作了解。

类似于Lettuce的配置，先引入Jedis依赖，配置如下：

```xml
<dependencies>
  <dependency>
    <groupId>redis.clients</groupId>
    <artifactId>jedis</artifactId>
    <version>3.6.0-RC1</version>
  </dependency>
</dependencies>
```

通过配置类向容器中添加连接器工厂，配置类代码如下：

```
@Configuration
class RedisConfiguration {
  @Bean
  public JedisConnectionFactory redisConnectionFactory() {
    RedisStandaloneConfiguration config = new RedisStandaloneConfiguration
("server", 6379);
    return new JedisConnectionFactory(config);
  }
}
```

5.1.2　RedisTemplate 与数据操作类的使用

类似于熟知的JdbcTemplate，Spring Data Redis提供了对Redis操作的高度抽象模板——RedisTemplate。RedisTemplate这个类在org.springframework.data.redis.core包下，是用来操作Redis最常用的核心类。

1．数据类型操作类的使用

Spring Data Redis对数据操作大致按类型来封装，对于不同的数据类型提供了对应的操作类，命名为xxxOperations。

这些数据类型操作类有GeoOperations、HashOperations、HyperLogLogOperations、ListOperations、SetOperations、ValueOperations和ZsetOperations等，每个Options类又有一个对应绑定Key（Key Bound）的操作类，具体是BoundGeoOperations、BoundHashOperations、BoundKeyOperations、BoundListOperations、BoundSetOperations、BoundValueOperations和BoundZSetOperations。绑定Key（Key Bound）的操作类其实就是在操作前为实例指定Key，在操作时不用每次都指定Key。

使用RedisTemplate的思路是先获取操作类实例。由于通过RedisTemplate获取Operations类实例十分频繁，Spring容器为注入做了优化。Spring容器通过调用方法opsFor[X]可以为Operations类型实例注入RedisTemplate实例，比如在容器中注入id为redisTemplate的RedisTemplate实例，代码如下：

```
@Bean
public RedisTemplate redisTemplate(RedisConnectionFactory
redisConnectionFactory){
    RedisTemplate redisTemplate = new RedisTemplate();
    redisTemplate.setConnectionFactory(redisConnectionFactory);
    return redisTemplate;
}
```

然后在使用时，通过注解@Resource和使用Operations类型变量来接收，代码如下：

```
@Resource(name = "redisTemplate")
ValueOperations ValueOperations;
```

类型自动转换使用起来十分方便。另外，由于java.lang.String类型使用的十分频繁，Spring Data Redis提供了用于处理String键和值的Template类——StringRedisTemplate。

2．序列化器的使用

StringRedisTemplate类继承自RedisTemplate，并指定了序列化器为StringRedisSerializer以及编码集为UTF-8。这样一来，使用StringRedisTemplate存储的字符串数据不用翻译即可直接被识别。

关于序列化器的使用有必要做一些说明。如果上面创建RedisTemplate时没有指定序列化器，那么Spring会采用默认的序列化器JdkSerializationRedisSerializer，这样一来，即使存储的是英文，在文本前面也会出现如"\xac\xed\x00\x05t\x00\"这样的字符，具体看下面的示例。

【示例5.2】 不指定序列化器时生成前缀字符

为容器配置RedisTemplate实例，未指定序列化器，代码如下：

```
@Bean
public LettuceConnectionFactory redisConnectionFactory() {
    return new LettuceConnectionFactory(new RedisStandaloneConfiguration
("localhost", 6379));
}
@Bean
public RedisTemplate redisTemplate(RedisConnectionFactory
redisConnectionFactory){
    RedisTemplate redisTemplate = new RedisTemplate();
    redisTemplate.setConnectionFactory(redisConnectionFactory);
    return redisTemplate;
}
```

在测试类中使用ValueOperations向Redis插入数据，key和value都是纯英文，代码如下：

```
@Resource(name = "redisTemplate")
ValueOperations ValueOperations;
@Test
void testOperationsRedisTemplate(){
    ValueOperations.set("ThisIsKey","This is value");
}
```

运行测试方法testOperationsRedisTemplate，通过redis-cli查看保存的数据，如图5.2所示。

在我们要保存的ThisIsKey之前的"\xac\xed\x00\x05t\x00\t"就是JDK的序列化器JdkSerializationRedisSerializer生成的。

解决这个问题的方法有两种：

（1）通过编码为RedisTemplate设置序列化器，代码如下：

```
redisTemplate.setKeySerializer( RedisSerializer.string());
redisTemplate.setValueSerializer( RedisSerializer.string());
```

（2）注入使用StringRedisTemplate，首先在容器中配置StringRedisTemplate，代码如下：

```
@Bean
public StringRedisTemplate stringRedisTemplate(RedisConnectionFactory
redisConnectionFactory){
    StringRedisTemplate stringRedisTemplate = new StringRedisTemplate();
    stringRedisTemplate.setConnectionFactory(redisConnectionFactory);
    return stringRedisTemplate;
}
```

然后在注入ValueOperations的位置指定stringRedisTemplate，代码如下：

```
@Resource(name = "stringRedisTemplate")
ValueOperations ValueOperations;
```

两种方法任选其一，最后再次运行测试方法testOperationsRedisTemplate，通过redis-cli查看结果，如图5.3所示，新增数据没有前缀了。

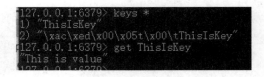

图 5.2　保存到 Redis 中的数据　　　　图 5.3　查看 Redis 中所有 key

5.1.3　RedisCallback、SessionCallback 接口和 Redis 事务的使用

RedisCallback、SessionCallback 接口是 Spring Data Redis 在 org.springframework.data.redis.core 包下提供的两个接口，这两个接口中都提供了一个抽象方法，在使用上也有相似之处。

RedisCallback 接口的代码如下：

```
package org.springframework.data.redis.core;

import org.springframework.dao.DataAccessException;
import org.springframework.data.redis.connection.RedisConnection;
import org.springframework.lang.Nullable;

public interface RedisCallback<T> {
    T doInRedis(RedisConnection connection) throws DataAccessException;
}
```

这个接口使用起来十分方便，只需要将接口实例传递给 RedisTemplate 的 execute 方法即可。通过方法的 connection 参数，我们能够使用相比于 RedisTemplate 更为底层的 RedisConnection。

【示例5.3】 使用 RedisCallback 查询数据

在开始之前已经向 Redis 中插入了一条数据，key 是"这是Key"，value 是"这是Value"。实现根据 key 查询 value 的功能，并在控制台输出 value。

使用的是 RedisTemplate 的子类 StringRedisTemplate，并使用拉姆达表达式创建接口实例，具体代码如下：

```
StringRedisTemplate stringRedisTemplate;
@Test
void testRedisCallback(){
    String value = stringRedisTemplate.execute((RedisCallback<String>)(connection) -> {
        byte[] valueBytes = connection.get("这是Key".getBytes(StandardCharsets.UTF_8));
        return new String(valueBytes, StandardCharsets.UTF_8);
    });
    System.out.println("value:"+value);
}
```

执行测试方法，控制台输出如图5.4所示。

图 5.4　使用 RedisCallback 查询数据日志输出

另一个接口SessionCallback在使用上和接口RedisCallback类似，同样可以通过调用RedisTemplate的execute方法来使用。这个接口的意义在于对Redis事务的支持。

Redis事务和传统数据事务并不相同，只能简单地理解成有顺序地执行多条命令。并且，Redis不支持回滚，所以在一次执行多条命令时，不能保证多条命令的原子性，也就是其中一条命令执行失败，不会使得其他命令回滚。

SessionCallback接口与RedisCallback接口的不同之处在于前者的方法定义，代码如下：

```
<K, V> T execute(RedisOperations<K, V> operations) throws DataAccessException;
```

该方法的参数是RedisOperations类型的，通过参数的multi和exec等方法可以方便地使用Redis事务功能。SessionCallback的execute方法的返回值也会作为RedisTemplate的execute方法的返回值返回。

【示例5.4】 使用SessionCallback执行多条命令

通过SessionCallback接口执行两个操作，在Redis中保存一个字符串和一个列表，具体代码如下：

```
@Test
void testSessionCallback(){
    List<Object> retList = stringRedisTemplate.execute(new SessionCallback<List<Object>>() {
        public List<Object> execute(RedisOperations operations) throws DataAccessException {
            operations.multi();
            operations.opsForValue().set("这的确是Key", "这还是Value");
            operations.opsForList().leftPushAll("设置一个List", "a", "b", "c");
            return operations.exec();
        }
    });
    for (Object o : retList) {
        System.out.println(o);
    }
}
```

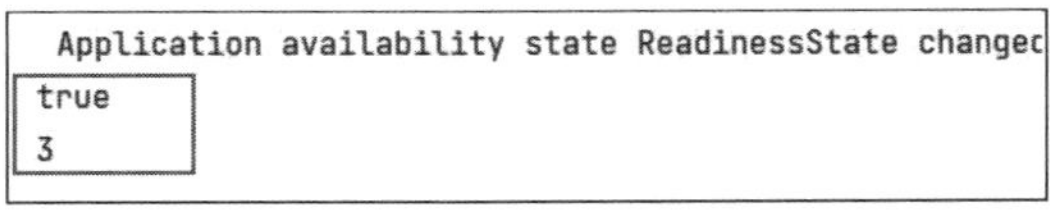

运行测试方法，执行结果如图5.5所示。

图 5.5 测试的输出结果

返回值是一个列表，每一项对应每个命令的执行结果。

5.2 在Spring Boot中配置和使用Redis

上一节介绍了Spring Data Redis项目，本节将介绍如何在Spring Boot项目中配置和使用Redis，主要是在Spring Boot中引入Spring Boot官方提供的spring-boot-starter-data-redis来操作Redis，以及进行Redis参数的配置。

5.2.1　通过 Starter 引入 Redis 相关依赖并配置 Redis

引入spring-boot-starter-data-redis和其他Starter一样非常简单，就是在POM文件中引入依赖，具体代码如下：

```xml
<dependency>
    <groupId>org.springframework.boot</groupId>
    <artifactId>spring-boot-starter-data-redis</artifactId>
</dependency>
```

查看spring-boot-starter-data-redis的POM文件，可以看到如下代码：

```xml
<dependency>
    <groupId>org.springframework.data</groupId>
    <artifactId>spring-data-redis</artifactId>
    <version>2.5.1</version>
    <scope>compile</scope>
</dependency>
<dependency>
    <groupId>io.lettuce</groupId>
    <artifactId>lettuce-core</artifactId>
    <version>6.1.2.RELEASE</version>
    <scope>compile</scope>
</dependency>
```

说明我们在引入这个Starter时，相当于间接地引入了Spring Data Redis项目和Lettuce连接器，因此上一节的知识在后面的开发中都可以使用到。

从配置中可以看出，Spring Boot的Starter默认采用了Lettuce连接器。如果必要，可以通过配置切换成Jedis，具体配置如下：

```xml
<dependency>
    <groupId>org.springframework.boot</groupId>
    <artifactId>spring-boot-starter-data-redis</artifactId>
    <exclusions>
        <exclusion>
            <groupId>io.lettuce</groupId>
            <artifactId>lettuce-core</artifactId>
        </exclusion>
    </exclusions>
</dependency>
<dependency>
    <groupId>redis.clients</groupId>
    <artifactId>jedis</artifactId>
</dependency>
```

在RedisProperties这个类中定义了可以在Spring Boot配置文件中使用的参数及其自动配置，具体列举在表5.1中。

表 5.1　Redis 配置文件

名　称	默 认 值	描　述
spring.redis.client-name		相当于使用Client Setname命令指定当前连接的名称
spring.redis.client-type		客户端也就是连接器的类型，可选ClientType.LETTUCE和ClientType.JEDIS。默认根据类路径下的类来检测
spring.redis.cluster.max-redirects		跨集群执行命令时要遵循的最大重定向数
spring.redis.cluster.nodes		以逗号分隔的"主机:端口"对列表进行引导。这表示集群节点的"初始"列表，并且要求至少有一个条目
spring.redis.connect-timeout		连接超时
spring.redis.database	0	连接的数据库索引
spring.redis.host	localhost	Redis服务器地址
spring.redis.jedis.pool.max-active	8	使用Jedis连接器时，连接池的最大连接数。使用负值表示没有限制
spring.redis.jedis.pool.max-idle	8	使用Jedis连接器时，连接池的最大空闲连接数。使用负值表示不限制
spring.redis.jedis.pool.max-wait	-1ms	使用Jedis连接器时，当池耗尽时，在抛出异常之前，连接分配应该阻塞的最大时间。使用负值表示无限期阻塞
spring.redis.jedis.pool.min-idle	0	使用Jedis连接器时，在连接池中维护的最小空闲连接数的目标
spring.redis.jedis.pool.time-between-eviction-runs		使用Jedis连接器时，空闲连接清理器线程两次运行之间的时间间隔。只有当为正值时，空闲连接清理器线程才执行，否则不执行
spring.redis.lettuce.cluster.refresh.adaptive	false	使用Lettuce连接器时，是否应该将所有可用的刷新触发器进行自适应拓扑刷新
spring.redis.lettuce.cluster.refresh.dynamic-refresh-sources	true	使用Lettuce连接器时，是否发现并查询集群的所有节点来获取集群的结构。当设置为false时，只有初始种子节点被用作拓扑发现源
spring.redis.lettuce.cluster.refresh.period		使用Lettuce连接器时，集群拓扑结构的刷新周期
spring.redis.lettuce.pool.max-active	8	使用Lettuce连接器时，连接池可以分配的最大连接数。负值表示不限制最大数量
spring.redis.lettuce.pool.max-idle	8	使用Lettuce连接器时，连接池的最大空闲连接数量
spring.redis.lettuce.pool.max-wait	-1	使用Lettuce连接器时，当池连接耗尽时，在抛出异常之前，连接分配应该阻塞的最大时间。负值表示阻塞时间没有限制
spring.redis.lettuce.pool.min-idle	0	使用Lettuce连接器时，连接池空闲时保留的最少连接数量
spring.redis.lettuce.pool.time-between-eviction-runs		使用Lettuce连接器时，空闲连接清理器线程之间运行的时间间隔。当为正值时，连接清理器线程启动，否则不执行
spring.redis.lettuce.shutdown-timeout	100ms	使用Lettuce连接器时，关闭超时时长
spring.redis.password		Redis Server的登录密码

（续表）

名　称	默　认　值	描　述
spring.redis.port		Redis Server的端口号
spring.redis.sentinel.master		Redis Server的名字
spring.redis.sentinel.nodes		以逗号分隔的host:port键值对列表
spring.redis.sentinel.password		Sentinel的授权密码
spring.redis.ssl	false	是否开启对SSL的支持
spring.redis.timeout		读取超时时长
spring.redis.url		连接到Redis Server的完整URL，格式为redis://user:password@host:port，其中的user会被忽略。该属性会覆盖host、port、password属性
spring.redis.username		登录Redis Server的属性名
spring.cache.redis.cache-null-values	true	是否允许缓存空值
spring.cache.redis.enable-statistics	false	是否开启缓存统计
spring.cache.redis.key-prefix		Key的统一前缀
spring.cache.redis.time-to-live		对象过期时长
spring.cache.redis.use-key-prefix	true	在向Redis中写入数据时是否使用key的前缀

其中关于cache的参数将会在5.4节中使用到。

5.2.2 Redis 数据类型及操作 API

学习Redis的使用主要是学习使用Redis提供的API，而在Redis中大部分API都归属于某一数据类型下，所以结合数据类型来学习Redis API是非常好的学习入口。

相比于早期的内存型数据库Memcached，Redis提供了丰富的数据类型。Redis总共提供了8种数据类型，分别有Strings、List、Sets、Sorted Sets、Hashes、Bit Arrays、HyperLogLogs和Streams。下面将一一介绍它们。

由于Redis是键值型存储的，每一条数据必有key，因此在讨论具体数据类型之前有必要对key做一定的说明。

1. key

Redis的key支持任何二进制序列和空值，所以任何二进制序列都可以用来作为数据的key。单个key最大大小为512MB，但是最好不要使用过长或过短的数据作为key。key过长会过多消耗不必要的内存、带宽和计算资源，而过短会牺牲数据的可读性，在实际工作中需要我们做好这两者的权衡。

对于key的操作，Redis提供了一系列的命令，列举在表5.2中。

表5.2 key 命令汇总

名　称	语　法	版　本	描　述
COPY	COPY source destination [DB destination-db] [REPLACE]	6.2.0	根据key复制数据，DB的索引是可选项，用来指定目标key的DB
DEL	DEL key1 key2 ...	1.0.0	删除指定的key，可以指定多个，会跳过不存在的key，返回删除key的个数

（续表）

名　　称	语　　法	版　　本	描　　述
DUMP	DUMP key	2.6.0	用来将指定key的数据序列化，RESTORE可以将序列化的结果反序列化到Redis中
EXISTS	EXISTS key [key ...]	1.0.0	从3.0.3版本起参数支持多个key，返回值存在key的个数，重复的key重复计数
EXPIRE	EXPIRE key seconds	1.0.0	指定或更新一个key的过期时间
EXPIREAT	EXPIREAT key timestamp	1.2.0	和EXPIRE命令类似，为key指定存活到的时间，参数为UNIX timestamp，单位是秒
EXPIRETIME	EXPIRETIME key	7.0.0	获取一个key的过期时间，返回值是过期时间的UNIX timestamp，单位也是秒。返回值-1表示key未指定过期时间，-2表示key不存在
KEYS	KEYS pattern	1.0.0	查找所有符合给定模式的key。支持?和*模糊匹配，使用反斜线处理特殊字符。返回值为keys名称列表
MIGRATE	MIGRATE host port key|"" destination-db timeout [COPY] [REPLACE] [AUTH password] [AUTH2 username password] [KEYS key [key ...]]	2.6.0	用户跨Redis实例的原子性数据迁移。原子性意味着一旦命令执行，会阻塞迁入和迁出两个Redis实例，直至迁移成功、失败或者超时。从3.0.3版本开始支持多个key，即可选项keys
MOVE	MOVE key db	1.0.0	将当前数据库的key移动到指定的db中。若key不存在或目标db中已经有了同名key，则不执行任何操作
OBJECT	OBJECT subcommand [arguments [arguments ...]]	2.2.3	允许从内部察看给定key的Redis对象，通常debug时使用。重点关注它的3个子命令，如下： OBJECT REFCOUNT \<key>返回指定key引用所存储的值的被引用次数。 OBJECT ENCODING \<key>返回给定key锁存储的值所使用的内部类型（Representation）。 OBJECT IDLETIME \<key>返回给定key的空转时间（Idle，没有被读取，也没有被写入），单位为秒
PERSIST	PERSIST key	2.2.0	清除指定key的过期时间。返回1表示正常清除，返回0表示key不存在或者key不存在过期时间
PEXPIRE	PEXPIRE key milliseconds	2.6.0	和EXPIRE几乎一样，只是参数以毫秒为单位
PEXPIREAT	PEXPIREAT key milliseconds-timestamp	2.6.0	和EXPIREAT几乎一样，只是时间戳以毫秒为单位
PEXPIRETIME	PEXPIRETIME key	7.0.0	和EXPIRETIME几乎一样，只是返回值单位为毫秒
PTTL	PTTL key	2.6.0	以毫秒为单位返回key的剩余生存时间

（续表）

名　称	语　法	版　本	描　述
RANDOMKEY	RANDOMKEY	1.0.0	返回当前数据库中的一个随机key，若数据库为空，则返回nil
RENAME	RENAME key newkey	1.0.0	为指定的key修改名字，改名成功时返回OK，失败时返回错误
RENAMENX	RENAMENX key newkey	1.0.0	与RENAME功能一致，只是在返回值上不同，改名成功时返回1，key的新名字已经存在时返回0
RESTORE	RESTORE key ttl serialized-value [REPLACE] [ABSTTL] [IDLETIME seconds] [FREQ frequency]	2.6.0	将给定的序列化的值反序列化，与DUMP是一对操作
SCAN	SCAN cursor [MATCH pattern] [COUNT count] [TYPE type]	2.8.0	该命令用于增量式迭代当前DB中的key，每次返回都会返回cursor，为下次传递的cursor值，当返回cursor为0时，表示已遍历到集合的末尾
SORT	SORT key [BY pattern] [LIMIT offset count] [GET pattern [GET pattern ...]] [ASC|DESC] [ALPHA] [STORE destination]	1.0.0	返回或保存给定Lists、Sets、Sorted Sets中经过排序的元素。排序默认以数字作为对象，值被解释为双精度浮点数，然后进行比较。SORT命令有丰富的参数列表，排序也非常灵活
TOUCH	TOUCH key [key ...]	3.2.1	更新key的最新访问时间
TTL	TTL key	1.0.0	与PTTL类似，以秒为单位返回key的剩余生存时间
TYPE	TYPE key	1.0.0	以字符串的形式返回数据类型，有String、List、Set、ZSet、Hash和Stream，当key不存在时返回none
UNLINK	UNLINK key [key ...]	4.0.0	和DEL功能一样，只是UNLINK不会立即回收key的Value的内容，因此也不会阻塞服务
WAIT	WAIT numreplicas timeout	3.0.0	用于阻塞当前客户端，直到所有先前写入的命令成功传输，在达到指定数量的副本或超时时，该命令将始终返回确认在WAIT命令之前发送的写命令的副本数量

2．Strings

Strings是Redis支持的最简单的类型，包含但并不限于常规的字符串，支持二进制数据。在官方文档中特别指出，Strings的值是二进制安全的。而所谓的二进制安全，是通过记录值的长度来判断值的结束的，而不是根据特殊符号来判断值的结束的，比如"\n"这样的特殊字符。通过特殊符号来判断值的结束是非二进制安全的。

Strings类型的操作命令列举在表5.3中。

表 5.3 Strings 命令汇总

名称	语法	版本	描述
APPEND	APPEND key value	2.0.0	在指定的key值末尾拼接value，如果key不存在，那么先创建一个值为空字符串的key，然后在末尾拼接
DECR	DECR key	1.0.0	将key中存储的数字值减一，当值不是数值类型时将返回错误，并且此操作限制在64位有符号整数范围内
DECRBY	DECRBY key decrement	1.0.0	将key所存储的值减去decrement
GET	GET key	1.0.0	获取key对应的值，如果key不存在，则返回nil；如果key存储的不是Strings类型，则返回错误
GETDEL	GETDEL key	6.2.0	获取key的值，并删除key
GETEX	GETEX key [EX seconds|PX milliseconds|EXAT timestamp|PXAT milliseconds-timestamp|PERSIST]	6.2.0	获取key的值，并且设置过期时间或移除过期时间。就像是把GET命令和EXPIRE、EXPIREAT、PEXPIRE、PEXPIREAT、PERSIST命令整合到一起使用
GETRANGE	GETRANGE key start end	2.4.0	这个命令在2.0版本之间叫SUBSTR。返回字符串包含start和end的子串。0表示第一个字符，负值从末尾开始计算，例如−1表示最后一个字符
GETSET	GETSET key value	1.0.0	给key设置新值，并返回原先的值，如果原先的值不是Strings类型，则返回错误
INCR	INCR key	1.0.0	将key中存储的数字值加一，当值不是数值类型时将返回错误，并且此操作限制在64位有符号整数范围内
INCRBY	INCRBY key increment	1.0.0	将key所存储的值加上increment
INCRBYFLOAT	INCRBYFLOAT key increment	2.6.0	将key所存储的值加上浮点数值increment
MGET	MGET key [key ...]	1.0.0	返回指定key的值，如果不存在则返回nil
MSET	MSET key value [key value ...]	1.0.1	批量设置key和value。对于已存在的key会覆盖
MSETNX	MSETNX key value [key value ...]	1.0.1	功能与MSET一致，只是在key已存在时不会覆盖
PSETEX	PSETEX key milliseconds value	2.6.0	PSETEX的工作方式与SETEX完全相同，唯一的区别是过期时间是以毫秒而不是以秒为单位指定的
SET	SET key value [EX seconds|PX milliseconds|EXAT timestamp|PXAT milliseconds-timestamp|KEEPTTL] [NX|XX] [GET]	1.0.0	给key设置Strings类型的值，如果key之前存在值，则会覆盖，并且原先的值可以是任何类型。若之前存在超时设置，则在SET命令执行成功后超时设置会失效
SETEX	SETEX key seconds value	2.0.0	这个命令是SET和EXPIRE的合并，不过是原子操作

（续表）

名称	语法	版本	描述
SETNX	SETNX key value	1.0.0	等价于SET命令，只是在key已存在时不会覆盖原先的值
SETRANGE	SETRANGE key offset value	2.2.0	将原先的字符串从偏移量offset开始覆盖上value内容。若key不存在，则当作空字符串处理。原字符串不够offset的部分会用零字节来处理"\x00"
STRALGO	STRALGO LCS algo-specific-argument [algo-specific-argument ...]	6.0.0	STRALGO实现了对字符串进行操作的复杂算法。当前唯一实现的算法是LCS（Longest Common Subsequence，最长公共子序列）算法
STRLEN	STRLEN key	2.2.0	返回字符串值的长度

【示例5.5】 给非数字值使用INCR命令

先给key"letter"设置一个字符串"a"，然后使用命令查看key"letter"的值，再执行INCR命令，在Redis客户端中的执行结果如图5.6所示。

```
127.0.0.1:6379> set letter a
OK
127.0.0.1:6379> get letter
a
127.0.0.1:6379> incr letter
ERR value is not an integer or out of range
```

图 5.6 测试 INCR 返回错误

3．Lists

Redis的Lists类型就是根据插入顺序排序的字符串集合，其基本实现形式是链表。Redis对于Lists提供的命令如表5.4所示。

表 5.4 Lists 命令汇总

名称	语法	版本	描述
BLMOVE	BLMOVE source destination LEFT|RIGHT LEFT|RIGHT timeout	6.2.0	是LMOVE命令的阻塞版本，与LMOVE不同的是，当source为空时会阻塞当前连接，直到其他客户端向source中插入数据或者连接超时。若timeout为0，则会无限等待下去
BLPOP	BLPOP key [key ...] timeout	2.2.0	是BLPOP命令的阻塞版本，与BLPOP不同的是，当key不存在时会阻塞当前连接，直到其他客户端向key中插入数据或者连接超时。若timeout为0，则会无限等待下去
BRPOP	BRPOP key [key ...] timeout	2.0.0	是RPOP的阻塞版本，其阻塞方式与BLPOP相似
BRPOPLPUSH	BRPOPLPUSH source destination timeout	2.2.0	是RPOPLPUSH的阻塞版本
LINDEX	LINDEX key index	1.0.0	返回index索引的元素，整数从0开始，–1表示最后一个元素。当index超过列表长度时返回nil
LINSERT	LINSERT key BEFORE|AFTER pivot element	2.2.0	在pivot之前（BEFORE）或之后（AFTER）插入element。当key不存在时，会当作空列表来处理。若列表中不存在pivot，则会返回错误

（续表）

名称	语法	版本	描述
LLEN	LLEN key	1.0.0	返回列表的长度，如果key不存在，则返回0。如果key非列表类型，则返回错误
LMOVE	LMOVE source destination LEFT|RIGHT LEFT|RIGHT	6.2.0	将source左边或右边的元素删除，并将删除的元素追加到destination左边或右边。这里左边为头，右边为尾
LPOP	LPOP key [count]	1.0.0	删除并返回存储在键值中的列表开头的count个元素，默认count为1
LPOS	LPOS key element [RANK rank] [COUNT num-matches] [MAXLEN len]	6.0.6	返回Redis列表中匹配元素的索引，从0开始。如果没有匹配到，那么返回nil
LPUSH	LPUSH key element [element ...]	1.0.0	插入所有指定的值到列表的头部。当指定多个element时，由于按顺序插入列表，因此在命令中最左侧的元素在Redis中位于列表的末尾
LPUSHX	LPUSHX key element [element ...]	2.2.0	和LPUSH一致，只是当key不存在时不执行操作
LRANGE	LRANGE key start stop	1.0.0	返回列表的start到stop的所有元素，偏移量从0开始，−1表示最后一个元素
LREM	LREM key count element	1.0.0	删除列表中的element元素，count有3种取值，如下： • count=0表示删除列表中所有与element相等的元素。 • count>0表示从列表头开始计数，移除count个与element相等的元素。 • count<0表示从列表尾开始计数，移除count个与element相等的元素
LSET	LSET key index element	1.0.0	将列表中索引为index的元素设置为element。如果索引超出列表范围，则返回错误
LTRIM	LTRIM key start stop	1.0.0	删除列表索引start至stop以外的元素，start和stop所在的元素将会被保留
RPOP	RPOP key [count]	1.0.0	移除并返回列表的最后一个元素。count参数是在6.2.0版本加入的
RPOPLPUSH	RPOPLPUSH source destination	1.2.0	这是一个原子性的操作，首先将source的最后一个元素移除，并将这个移除的元素追加到destination的开头
RPUSH	RPUSH key element [element ...]	1.0.0	将指定的元素插入list的末尾。从2.4版本开始支持多个element
RPUSHX	RPUSHX key element [element ...]	2.2.0	与RPUSH基本一致，只是RPUSHX命令在key不存在时不执行操作

4. Sets

Sets是元素唯一但无序的Strings元素集合。Redis对于Sets提供的命令如表5.5所示。

表 5.5 Sets 命令汇总

名 称	语 法	版 本	描 述
SADD	SADD key member [member ...]	1.0.0	向key中添加一个或多个member，当key不存在时创建一个set，当member已存在时忽略。从2.4版本开始支持多个member
SCARD	SCARD key	1.0.0	返回set中元素的个数
SDIFF	SDIFF key [key ...]	1.0.0	以列表形式返回在第一个key中却不在第二个key及以后的key中的元素
SDIFFSTORE	SDIFFSTORE destination key [key ...]	1.0.0	与SDIFF命令类似，只是将SDIFF命令的结果保存到destination中。如果destination已存在，则会覆盖已有内容，返回结果集合中的元素个数
SINTER	SINTER key [key ...]	1.0.0	返回所有集合的交集元素
SINTERSTORE	SINTERSTORE destination key [key ...]	1.0.0	与SINTER命令类似，只是将SINTER命令的结果保存到destination中。如果destination已存在，则会覆盖已有内容，返回结果集合中的元素个数
SISMEMBER	SISMEMBER key member	1.0.0	如果member是key的成员，则返回1；如果不是或者key不存在，则返回0
SMEMBERS	SMEMBERS key	1.0.0	返回key的所有成员
SMISMEMBER	SMISMEMBER key member [member ...]	6.2.0	和SISMEMBER功能类似，只是支持同时检测多个member
SMOVE	SMOVE source destination member	1.0.0	将member从source中移动到destination中，相比于我们两次执行命令，这个命令的好处是它是原子操作
SPOP	SPOP key [count]	1.0.0	从key中随机移除并返回count个元素，count默认为1。从3.2.0版本开始支持count参数
SRANDMEMBER	SRANDMEMBER key [count]	1.0.0	从key中随机获取count个元素，count默认为1
SREM	SREM key member [member ...]	1.0.0	从key中移除指定的member，如果指定的member在key中不存在，则忽略，返回从key中移除的个数
SSCAN	SSCAN key cursor [MATCH pattern] [COUNT count]	2.8.0	用来遍历Sets类型中的元素，使用方法和SCAN一致
SUNION	SUNION key [key ...]	1.0.0	对指定的集合取并集，并返回结果
SUNIONSTORE	SUNIONSTORE destination key [key ...]	1.0.0	与SUNION命令类似，只是将SUNION命令的结果保存到destination中。如果destination已存在，则会覆盖已有内容，返回结果集合中的元素个数

5. Sorted Sets

和Sets非常相似，其元素都是唯一的，但是不同的是Sorted Sets是有序集合，其中的元素都有一个浮点类型的score值，Sorted Sets中的元素也根据score值来排序。

Redis为Sorted Sets提供的命令是所有数据类型中最多的，具体命令列在表5.6中。

表 5.6　Sorted sets 命令汇总

名　称	语　法	版　本	描　述
BZPOPMAX	BZPOPMAX key [key ...] timeout	5.0.0	是ZPOPMAX的阻塞版本。在指定的集合中没有数据时会阻塞连接，直至有数据或超时
BZPOPMIN	BZPOPMIN key [key ...] timeout	5.0.0	是ZPOPMIN的阻塞版本。在指定的集合中没有数据时会阻塞连接，直至有数据或超时
ZADD	ZADD key [NX|XX] [GT|LT] [CH] [INCR] score member [score member ...]	1.2.0	将指定的member和score添加到key中。如果member已存在，则更新已有的score值，并将member放到应该存在的位置。另外，有几个可选参数如下： • XX：仅更新，不添加新元素。 • NX：仅添加新元素，不更新。 • LT：对于更新操作，仅当新score小于当前score时更新。 • GT：对于更新操作，仅当新score大于当前score时更新。 • CH：返回的结果值包含修改的元素个数。 • INCR：指定此参数后，该命令类似于ZINCRBY命令
ZCARD	ZCARD key	1.2.0	返回存储的元素个数
ZCOUNT	ZCOUNT key min max	2.0.0	返回score值在min和max之间的元素个数
ZDIFF	ZDIFF numkeys key [key ...] [WITHSCORES]	6.2.0	计算第一个和所有后续的set之间的不同元素，并将结果返回。输入键的总数由numkeys指定。可选WITHSCORES一并返回元素的score值
ZDIFFSTORE	ZDIFFSTORE destination numkeys key [key ...]	6.2.0	计算第一个和所有后续的set之间的不同元素，并将结果存储在destination中。输入键的总数由numkeys指定。可选WITHSCORES一并返回元素的score值

（续表）

名称	语法	版本	描述
ZINCRBY	ZINCRBY key increment member	1.2.0	给指定的member的score值增加increment（当increment为负值时就是减少）。如果key不存在，则创建新的set；如果member不在key中，则将member添加到key中，并且increment作为其score值
ZINTER	ZINTER numkeys key [key ...] [WEIGHTS weight [weight ...]] [AGGREGATE SUM|MIN|MAX] [WITHSCORES]	6.2.0	计算指定集合的交集，并将交集的元素返回
ZINTERSTORE	ZINTERSTORE destination numkeys key [key ...] [WEIGHTS weight [weight ...]] [AGGREGATE SUM|MIN|MAX]	2.0.0	计算指定集合的交集，并将其元素与所有score之和存储到destination中
ZLEXCOUNT	ZLEXCOUNT key min max	2.8.9	获取指定的字典排序范围中的元素个数
ZMSCORE	ZMSCORE key member [member ...]	6.2.0	返回member的score值
ZPOPMAX	ZPOPMAX key [count]	5.0.0	返回并移除Sorted Sets中score值最大的count个元素
ZPOPMIN	ZPOPMIN key [count]	5.0.0	返回并移除Sorted Sets中score值最小的count个元素
ZRANDMEMBER	ZRANDMEMBER key [count [WITHSCORES]]	6.2.0	若count为正数，且count小于等于集合的大小，则返回count个不相同的元素，若count大于集合的大小，则返回集合所有元素；若count为负数，则返回|count|个元素，允许相同元素存在
ZRANGE	ZRANGE key min max [BYSCORE|BYLEX] [REV] [LIMIT offset count] [WITHSCORES]	1.2.0	返回指定范围的元素。这里范围默认是索引，可选score值和字典排序。元素的顺序按score值从低到高排列，得分相同的元素按字典顺序排列。可选参数REV，将结果集的排列顺序颠倒
ZRANGEBYLEX	ZRANGEBYLEX key min max [LIMIT offset count]	2.8.9	在版本6.2.0开始被标记为过期，推荐使用携带BYLEX参数的ZRANGE命令
ZRANGEBYSCORE	ZRANGEBYSCORE key min max [WITHSCORES] [LIMIT offset count]	1.0.5	在版本6.2.0开始被标记为过期，推荐使用携带BYSCORE参数的ZRANGE命令

（续表）

名　称	语　法	版　本	描　述
ZRANGESTORE	ZRANGESTORE dst src min max [BYSCORE|BYLEX] [REV] [LIMIT offset count]	6.2.0	该命令是将ZRANGE的结果集存储到dst中，并返回结果集的个数
ZRANK	ZRANK key member	2.0.0	返回member在集合中的排名，排名列表按score值从小到大排列，并且排名是从0开始的
ZREM	ZREM key member [member ...]	1.2.0	从集合中移除一个或多个member，返回实际删除的元素个数
ZREMRANGEBYLEX	ZREMRANGEBYLEX key min max	2.8.9	根据字典排序的范围删除元素
ZREMRANGEBYRANK	ZREMRANGEBYRANK key start stop	2.0.0	根据默认排序的范围删除元素
ZREMRANGEBYSCORE	ZREMRANGEBYSCORE key min max	1.2.0	根据score值的范围删除元素
ZREVRANGE	ZREVRANGE key start stop [WITHSCORES]	1.2.0	和ZRANGE命令类似，只是降序排列
ZREVRANGEBYLEX	ZREVRANGEBYLEX key max min [LIMIT offset count]	2.8.9	和ZRANGEBYLEX命令类似，只是按字典顺序降序排列
ZREVRANGEBYSCORE	ZREVRANGEBYSCORE key max min [WITHSCORES] [LIMIT offset count]	2.2.0	和ZRANGEBYSCORE命令类似，只是按SCORE值降序排列
ZREVRANK	ZREVRANK key member	2.0.0	获取member在降序排列中的排名，排名从0开始
ZSCAN	ZSCAN key cursor [MATCH pattern] [COUNT count]	2.8.0	和SCAN的用法一致
ZSCORE	ZSCORE key member	1.2.0	返回member的score值
ZUNION	ZUNION numkeys key [key ...] [WEIGHTS weight [weight ...]] [AGGREGATE SUM|MIN|MAX] [WITHSCORES]	6.2.0	计算多个集合的并集，并将并集返回，其score值是元素在所有集合中score值的和
ZUNIONSTORE	ZUNIONSTORE destination numkeys key [key ...] [WEIGHTS weight [weight ...]] [AGGREGATE SUM|MIN|MAX]	2.0.0	计算多个集合的并集，并将并集保存到destination中，其score值是元素在所有集合中score值的和

6. Hashes

Hashes相当于存储键值型的字典，其内部有field和value的概念，field和value都是Strings类型的数据。Redis提供的Hashes相关的命令列举在表5.7中。

表 5.7 Hashes 命令汇总

名　称	语　法	版　本	描　述
HDEL	HDEL key field [field ...]	2.0.0	删除Hashes中的field，从2.4版本后支持多个field。当key或field不存在时不会返回错误，返回值返回实际删除的field个数
HEXISTS	HEXISTS key field	2.0.0	检查是否存在指定的字段，如果存在则返回1，否则返回0
HGET	HGET key field	2.0.0	获取field绑定的value，如果field或key不存在，则返回nil
HGETALL	HGETALL key	2.0.0	获取Hashes对象的所有数据，返回结果为list，由field和value依次衔接
HINCRBY	HINCRBY key field increment	2.0.0	给指定的field绑定的value增加increment
HINCRBYFLOAT	HINCRBYFLOAT key field increment	2.6.0	给指定的field绑定的value增加浮点类型的increment
HKEYS	HKEYS key	2.0.0	返回所有的field名称
HLEN	HLEN key	2.0.0	返回field的数量
HMGET	HMGET key field [field ...]	2.0.0	获取指定字段的值，如果字段不存在，则对应返回nil
HMSET	HMSET key field value [field value ...]	2.0.0	向字典中设置field和value，对于已经存在的field将会覆盖。这个命令在4.0.0版本中标记为过时，建议使用HSET命令来代替
HRANDFIELD	HRANDFIELD key [count [WITHVALUES]]	6.2.0	从Hashes中随机获取field，WITHVALUES可选项表示同时返回field的值。如果count为正，则field不重复，所以数量最大为HLEN的大小；如果count为负，则返回的field可能重复
HSCAN	HSCAN key cursor [MATCH pattern] [COUNT count]	2.8.0	用来遍历Hashes的命令，和SCAN一致
HSET	HSET key field value [field value ...]	2.0.0	向Hashes中设置field和value。从4.0.0版本开始支持一次设置多个field和value
HSETNX	HSETNX key field value	2.0.0	向Hashes中设置field和value，但只有当field不存在时才能设置成功
HSTRLEN	HSTRLEN key field	3.2.0	返回field关联的值的字符串长度
HVALS	HVALS key	2.0.0	返回Hashes中存储的所有value

7．Bit Arrays（Bitmaps）

Bit Arrays或Bitmaps并不能算是一种数据类型，而是Redis提供的一些命令。用这些命令可以将Strings数据按位操作，例如可以用来设置和清除单个位，计算所有设置为1的位，以及查找第一个设置或未设置的位，等等。具体命令展示在表5.8中。

表 5.8 Bit Arrays（Bitmaps）命令汇总

名 称	语 法	版 本	描 述
BITCOUNT	BITCOUNT key [start end]	2.6.0	计算给定的字符串或字符串的子串中位设置为1的数量
BITFIELD	BITFIELD key [GET type offset] [SET type offset value] [INCRBY type offset increment] [OVERFLOW WRAP|SAT|FAIL]	3.2.0	有3个子命令：GET、SET和INCRBY，分别是按位获取、设置以及增减操作。offset是偏移量，从0开始，偏移多少位开始计算。type由i和u拼接数字构成，i表示有符号整数，u表示无符号整数
BITOP	BITOP operation destkey key [key ...]	2.6.0	在多个key之间执行位操作，并将结果保存到destkey中。这些位操作包括AND、OR、XOR和NOT
BITPOS	BITPOS key bit [start] [end]	2.8.7	获取字符串或子串中第一个为bit的位的位置，bit取1或0
GETBIT	GETBIT key offset	2.2.0	获取在offset的那一位的bit值
SETBIT	SETBIT key offset value	2.2.0	设置或清除在offset位上的bit值

【示例5.6】 使用GETBIT命令查看字符串"a"的各位数值

在此之前已经将test_bit设置为字符串"a"，下面先用BITCOUNT命令查看字符串"a"中有多少位是1，命令如下：

```
127.0.0.1:6379> bitcount test_bit
3
```

上面输出为3，说明共有3个位是1。由于已知字符串"a"的二进制是"01100001"，因此我们可以直接获取第二位、第三位和第八位，代码如下：

```
127.0.0.1:6379> getbit test_bit 1
1
127.0.0.1:6379> getbit test_bit 2
1
127.0.0.1:6379> getbit test_bit 7
1
```

只有3个1，所以其余各位都为0，这与已知内容相符合。

8．HyperLogLogs

HyperLogLogs是一个用于估计集合的基数的概率数据结构。HyperLogLog算法在输入元素的数量或者体积非常大时，计算基数所需的空间总是固定的。在Redis中，每个HyperLogLog键只需要12KB的内存就可以计算接近2^64个不同元素的基数。Redis为HyperLogLogs提供了3个命令，具体列举在表5.9中。

表 5.9　HyperLogLogs 命令汇总

名　　称	语　　法	版　　本	描　　述
PFADD	PFADD key [element [element ...]]	2.8.9	将指定元素添加到HyperLogLog中
PFCOUNT	PFCOUNT key [key ...]	2.8.9	返回给定HyperLogLog的基数估算值
PFMERGE	PFMERGE destkey sourcekey [sourcekey ...]	2.8.9	将多个HyperLogLog合并为一个HyperLogLog

9．Streams

Streams是Redis 5.0.0版本开始支持的功能，主要用于实现消息队列，是对于发布订阅功能的补充。除了发布订阅实现的消息分发外，还可以实现消息的持久化和主备复制功能。在了解了Redis发布订阅功能之后，再来看Streams数据类型和其命令会更容易理解，发布订阅功能将在5.3.3节介绍。关于Streams的命令列举在表5.10中。

表 5.10　Streams 命令汇总

名　　称	语　　法	版　　本	描　　述
XACK	XACK key group ID [ID ...]	5.0.0	将消息标记为"已处理"
XADD	XADD key [NOMKSTREAM] [MAXLEN|MINID [=|~] threshold [LIMIT count]] *|ID field value [field value ...]	5.0.0	向队列追加消息，如果指定的队列不存在，则创建一个队列。ID可以使用星号来表示，由Redis来创建ID，这个ID值将会被返回
XDEL	XDEL key ID [ID ...]	5.0.0	根据ID来删除消息，支持多个ID
XGROUP	XGROUP [CREATE key groupname ID|$ [MKSTREAM]] [SETID key groupname ID|$] [DESTROY key groupname] [CREATECONSUMER key groupname consumername] [DELCONSUMER key groupname consumername]	5.0.0	这个命令有几个常用的子命令： • XGROUP CREATE：用来创建消费者组。 • XGROUP SETID：为消费者组设置新的最后传递的消息ID。 • XGROUP DELCONSUMER：删除消费者组中的消费者。 • XGROUP DESTROY：删除消费者组
XINFO	XINFO [CONSUMERS key groupname] [GROUPS key] [STREAM key] [HELP]	5.0.0	查看流和消费者组的相关信息，也有几个子命令： • XINFO GROUPS：查看消费者组的信息。 • XINFO STREAM：查看流的信息
XLEN	XLEN key	5.0.0	获取流包含的元素数量，即消息长度
XPENDING	XPENDING key group [[IDLE min-idle-time] start end count [consumer]]	5.0.0	显示待处理消息的相关信息
XRANGE	XRANGE key start end [COUNT count]	5.0.0	获取消息列表，通过start和end指定索引
XREAD	XREAD [COUNT count] [BLOCK milliseconds] STREAMS key [key ...] ID [ID ...]	5.0.0	以阻塞或非阻塞方式获取消息列表。其中的milliseconds参数，如果指定了，就是阻塞的时长；如果没有指定，就是不阻塞

（续表）

名称	语法	版本	描述
XREVRANGE	XREVRANGE key end start [COUNT count]	5.0.0	获取反向排序消息列表，顺序与XRANGE相反
XTRIM	XTRIM key MAXLEN|MINID [=|~] threshold [LIMIT count]	5.0.0	用来修剪队列。有两种指定方式，一是通过MAXLEN指定队列长度，当队列超过MAXLEN指定的长度时自动丢弃最早的数据；二是通过MINID指定最小ID，用来删除ID小于指定值的元素

以上是Redis提供的一些基本数据类型和相关的命令，此外还有GEO和发布订阅，发布订阅功能在5.3.3节中有介绍，不过对于GEO没有提及，如果需要，可以在Redis官网查看相关信息。

5.3 Redis的一些特殊用法

5.3.1 Redis 事务

如果熟悉关系型数据库，那么对"事务"一定再熟悉不过了。在Redis中也有事务的概念，但是与传统关系型数据库的"事务"有一些区别，在这里我们更多关注这些区别。在Redis 2.6.0版本引入Lua脚本后，Redis就有了两种实现事务的方式，原先就支持的Redis事务和Lua脚本，Lua脚本会在5.3.4节介绍。

Redis提供了5个事务相关的命令，分别是WATCH、UNWATCH、MULTI、EXEC和DISCARD。基本的使用流程就是先使用命令MULTI来开启事务，接下来客户端向Redis服务发送的命令不会被立即执行，而是被按顺序记录到一个队列中，最后当调用EXEC命令时，队列中的命令被依次执行。

上面就是Redis事务的基本使用。Redis事务这一机制可以提供两方面的保证：

- 一是队列中的命令在执行过程中不会被其他命令打断。
- 二是队列中的命令要么全部执行，要么全部不执行。

对于这两方面保证的描述，我们需要更加详细地体会。前一个保证是指在队列中的命令开始执行之后到执行完成不会被打断。注意这是在EXEC执行之后，而不是MULTI执行之后。后一个保证是说队列中的所有命令都会被执行，并不是值会被成功执行，例如队列中有多条命令，其中一条命令执行失败，并不会影响其他成功执行的命令。

注意　Redis不支持回滚操作。

执行命令使用EXEC，清空命令队列并放弃事务使用DISCARD，执行DISCARD之后，客户端从事务状态退出。

另外两个命令WATCH和UNWATCH是Redis对使用check-andset（CAS）操作实现乐观锁的支持。具体步骤是，在MULTI之前使用WATCH来监视一个或多个key，然后如同前面的操作发送命令，最后执行EXEC命令，Redis会检查被监视的key的值在执行EXEC之前是否发生了修改，如果是，则取消事务。

【示例5.7】 使用WATCH命令实现对数据的自增操作

执行Redis命令，监视key"mycount"，命令如下：

```
WATCH mycount
```

通过编程获取mycount的值，并对其加一，伪代码如下：

```
valueOfMycount = GET mycount
valueOfMycount = valueOfMycount + 1
```

下面执行Redis命令，其中$valueOfMycount表示增一后的值，命令如下：

```
MULTI
SET mycount $valueOfMycount
EXEC
```

上面的代码中，在EXEC执行之前，如果mycount的值发生变化，那么执行失败，EXEC将返回控制。我们可以在程序中反复执行上面的代码，直到程序没有发生碰撞为止。这就实现了通过乐观锁机制对数据的自增操作。当然这里只是示例，如果只是自增操作，我们可以简单地通过INCR命令来实现。

另一个命令UNWATCH用来取消对key的监视。不过，很多时候在使用EXEC或DISCARD时会自动取消之前的所有监视，也就不需要我们直接使用UNWATCH。

5.3.2 Redis Pipelined 和 Lua 脚本

Pipelined可以翻译为管道或流水线，是一种减少客户端与服务端网路通信等待时间的请求响应协议。使用Redis Pipelined，客户端可以为响应上一次请求向服务器发送下一条命令，这相比于等待响应后再发出下一条请求，尤其是在频繁向服务器发送命令时，大大提高了Redis服务的效率。

从Redis 2.6版本开始，通过内置Lua解释器支持Lua脚本。其使用起来非常简单，基本就是通过EVAL命令来执行，具体在这里不展开，如果需要，可以参考官方文档关于EVAL的介绍。

由于Lua脚本在写操作前无须等待读操作响应，因此相比于Pipelined，Lua脚本很多时候都更为高效。另外，Redis也支持在Pipelined中发送Lua脚本的EVAL和EVALSHA命令。

5.3.3 Redis 发布订阅

发布订阅是一种消息通信模式，由订阅者订阅频道，发布者向频道发送消息，Redis将消息发送给频道的订阅者。

订阅者（客户端）订阅通过SUBSCRIBE、PSUBSCRIBE来实现，一个订阅者可以订阅多个频道，一个频道也可以被多个订阅者订阅，大致结构如图5.7所示。

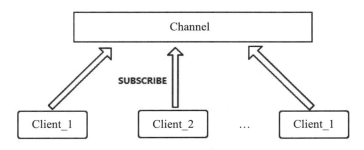

图 5.7　Redis 订阅的结构图

发布者（也是客户端）通过命令PUBLISH向频道发布消息，消息经由频道发送给订阅者，大致结构如图5.8所示。

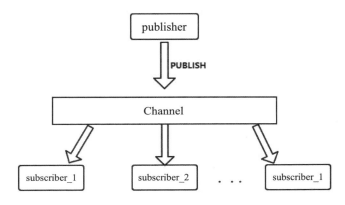

图 5.8　Redis 发布的结构图

5.4　使用Spring缓存注解操作Redis

Spring框架提供了缓存抽象，Spring Data Redis项目实现了这一抽象，具体实现在org.springframework.data.redis.cache包下。本节将介绍如何使用Spring提供的缓存注解来操作Redis。

5.4.1　启用缓存和配置缓存管理器

Spring Boot项目启用Spring缓存非常简单，只需要在程序启动类上添加@EnableCaching注解即可。其实在添加上@EnableCaching注解之后，程序已经可以使用缓存了，只是此时使用的是内存型缓存，默认的缓存管理器是ConcurrentMapCacheManager。可以使用如下代码查看缓存管理器：

```
ConfigurableApplicationContext context = SpringApplication.run
(MerchmanagerApplication.class, args);
CacheManager bean = context.getBean(CacheManager.class);
System.out.println(bean);
```

输出如下：

```
org.springframework.cache.concurrent.ConcurrentMapCacheManager@e09f1b6
```

在启用缓存之后，还可以自定义配置缓存管理器。因为我们要使用Spring Data Redis的缓存实现，所以需要配置的是Redis缓存管理器，通过代码引入程序中，具体代码如下：

```
@Bean
public RedisCacheManager cacheManager(RedisConnectionFactory
connectionFactory) {
    return RedisCacheManager.create(connectionFactory);
}
```

配置之后，从上面的输出可以看出容器中已经切换为Redis缓存管理器，具体如下：

```
org.springframework.data.redis.cache.RedisCacheManager@3724ab90
```

5.4.2 使用缓存注解开发

Spring Integration提供了缓存抽象注解，使用这些注解可以很方便地使用缓存功能。常用的缓存注解有两个，分别是@Cacheable和@CacheEvict。下面将分别介绍这两个注解。

1. @Cacheable

@Cacheable常用来标注在方法上，被标注的方法每次在被调用之前会先检查缓存中是否有值，如果没有缓存，则会调用方法，然后将结果缓存；如果有缓存，则会直接返回缓存的内容。

类似于Redis的key-value结构，缓存的创建和查询也是通过key来识别的。关于key的生成，最简单的形式是为@Cacheable指定默认参数，默认参数会作为key的一部分，比如如下参数：

```
@Cacheable("merch")
```

此时在Redis中生成的key如下：

```
merch::SimpleKey []
```

这里提及一个小细节，就是关于缓存的value，由于涉及序列化和反序列化，因此对象类型的返回值必须实现可序列化接口，不然会报错java.io.NotSerializableException。

在方法有形参列表时，默认情况下，所有形参都会生成在key中，如果我们不需要使用所有形参来区分缓存，可以使用该注解的key属性。key属性中支持SpEL表达式，所以可以非常灵活地使用形参或形参的属性值。下面是使用属性key的示例。

【示例5.8】 使用SpEL选取形参或属性

选中String类型的形参，代码如下：

```
@Cacheable(cacheNames="books", key="#author")
public Book findBook(String author, boolean readed)
```

选取形参的属性值，代码如下：

```
@Cacheable(cacheNames="books", key="#book.author")
public Book findBook(Book book, boolean readed)
```

调用某一类型的静态方法：

```
@Cacheable(cacheNames="books", key="T(someType).hash(#book)")
public Book findBook(Book book, boolean readed)
```

2. @CacheEvict

@CacheEvict和@Cacheable在作用上正好相反，用于移除缓存数据。不过在使用上有相似之处，比如默认的属性cacheNames和属性key，在使用上两个注解一致。

@CacheEvict更像是一个触发器，触发缓存删除的动作。这个触发时机有方法执行之前和执行之后可选，默认是方法执行之后。也可以通过配置beforeInvocation=true来设置为执行之前触发删除缓存。

对于删除缓存的指定，通常在删除时不需要具体的单个缓存，而是一类的删除，比如设置如下：

```
@CacheEvict(cacheNames="books", allEntries=true)
```

通过设置allEntries=true可以删除cacheNames="books"的一类缓存。

5.4.3 类实例方法类内部调用时的失效问题

由于缓存的是通过动态代理，通过接口实现的，所以类内部的方法互相调用时不会使用动态代理，代码如下：

```
@Override
public List<Merchandise> findAll() {
    System.out.println("MerchandiseService.findAll()");
    return finds();
}
@Override
@Cacheable("merch")
public List<Merchandise> finds(){
    System.out.println("MerchandiseService.finds()");
    return merchandiseDao.findAll();
}
```

在方法findAll()中调用了方法finds()，缓存注解标注在方法finds()上。此时，从外部调用方法findAll()，缓存机制便无法生效。

解决方法其实也简单，通过程序上下文对象获取当前实例对象，用该实例对象调用finds()。可以通过程序启动类获取上下文对象，代码如下：

```
@EnableCaching
@SpringBootApplication
public class MerchmanagerApplication {
    public static ApplicationContext context;
    public static void main(String[] args) {
        ConfigurableApplicationContext context = SpringApplication.run
(MerchmanagerApplication.class, args);
```

```
        MerchmanagerApplication.context = context;
    }
}
```

并修改findAll()方法，先获取当前类实例，然后通过实例调用finds()方法，代码如下：

```
@Override
public List<Merchandise> findAll() {
    System.out.println("MerchandiseService.findAll()");
    MerchandiseService merchandiseService = MerchmanagerApplication.context.getBean(MerchandiseService.class);
    System.out.println(merchandiseService.getClass());
    return merchandiseService.finds();
}
```

修改findAll()方法后，缓存就会生效。该方法中增加了一行输出实例merchandiseService的类型的方法，其输出如下：

```
class com.example.merchmanager.service.impl.MerchandiseServiceImpl$$EnhancerBySpringCGLIB$$9fd1f17e
```

可以看出这是一个代理实例，所以不难理解通过此实例缓存可以生效。

查看实例类型调用的是class属性，没有调用toString()方法，原因是代理类会将toString()代理给MerchandiseServiceImpl实例，所以使用toString()看不到代理实例。

5.4.4 缓存脏数据说明

缓存通常用于高并发的情况下，将频繁访问的数据存储到缓存中，从而降低数据库、磁盘的压力。使用缓存相当于同样的数据在两个系统中保存了，这时就难免出现数据不一致的情况。缓存就像真实数据的影子，当真实数据发生变化而缓存没有被修改时，这时缓存中的数据就是"脏数据"。

对于避免"脏数据"，我们通常可以在修改数据时通过@CacheEvict注解删除缓存。除了使用注解删除缓存数据外，也可以为缓存设置有效时长，当超过一定时间后自动删除缓存。

我们可以通过自定义Redis缓存管理器设置超时时间。自定义的缓存管理器只需创建一个类继承RedisCacheManager，并重写createCache()方法，在这个方法中指定存活时长。

5.5 实战——用Redis改版商品信息管理系统V2.0

本节将在4.5节项目的基础上对商品信息管理系统进行改造，引入Redis用作缓存。

在上一节中已经实现了商品和供应商的增、删、改，以及商品和供应商的两个列表。接下来，我们要做的就是在列表的Service接口中存入缓存，以及在增、删和改对应的接口中删除缓存。

5.5.1 引入 Redis 的依赖并配置 Redis 服务地址和启用缓存

在pom.xml中引入Redis的Starter，配置代码如下：

```xml
<dependency>
  <groupId>org.springframework.boot</groupId>
  <artifactId>spring-boot-starter-data-redis</artifactId>
</dependency>
```

在application.yaml中配置Redis的服务地址，配置代码如下：

```
spring:
  redis:
    host: "192.168.110.128"
```

上面配置了Redis服务的IP地址。

然后，在程序启动类MerchmanagerApplication上添加注解@EnableCaching以启用缓存，代码如下：

```
package com.example.merchmanager;

import org.springframework.boot.SpringApplication;
import org.springframework.boot.autoconfigure.SpringBootApplication;
import org.springframework.cache.CacheManager;
import org.springframework.cache.annotation.EnableCaching;
import org.springframework.context.ApplicationContext;
import org.springframework.context.ConfigurableApplicationContext;

@EnableCaching
@SpringBootApplication
public class MerchmanagerApplication {
    public static ApplicationContext context;
    public static void main(String[] args) {
        ConfigurableApplicationContext context = SpringApplication.run(MerchmanagerApplication.class, args);
        CacheManager bean = context.getBean(CacheManager.class);
        System.out.println(bean);
        MerchmanagerApplication.context = context;
    }
}
```

5.5.2 添加@Cacheable 和@CacheEvict 注解

以商品为例，查询列表的方法MerchandiseServiceImpl.findAll()，这个方法相当于查询列表，在这个方法上添加注解@Cacheable，这样每次访问该方法时，都会先在缓存中取数据，只有缓存中没有数据时才访问数据库，如此便减轻了数据库的访问压力，同时也加快了系统的响应速度。为了方便后面测试，增加一行输出，具体修改后的findAll()方法代码如下：

```
@Override
@Cacheable("merch")
public List<Merchandise> findAll() {
```

```
        System.out.println("MerchandiseService.findAll()");
        return merchandiseDao.findAll();
    }
```

注意，我们将缓存的key设置为merch。

上面是查询方法，向缓存中设置数据。为了使程序能够正确运行，接下来需要在数据发生变动时使用@CacheEvict从缓存中删除缓存。在我们的程序中，修改和创建使用了同一个方法MerchandiseServiceImpl.save(Merchandise entity)，删除使用了另一个方法MerchandiseServiceImpl.deleteById(Integer id)。所以我们需要在这两个方法上使用注解@CacheEvict，同时为方便测试增加了日志输出，具体代码如下：

```
    @Override
    @CacheEvict(cacheNames = "merch", allEntries = true)
    public Merchandise save(Merchandise entity) {
        System.out.println("MerchandiseService.save");
        return merchandiseDao.save(entity);
    }

    @Override
    @CacheEvict(cacheNames = "supp", allEntries = true)
    public void deleteById(Integer id) {
        System.out.println("MerchandiseService.deleteById");
        merchandiseDao.deleteById(id);
    }
```

这里同样为注解默认值指定了merch，这样才能保证上面的添加缓存和这里的删除缓存是对于同一条缓存。

对照商品管理的修改，供应商管理是同样的修改思路，具体代码如下：

```
package com.example.merchmanager.service.impl;

import com.example.merchmanager.dao.SupplierDao;
import com.example.merchmanager.pojo.Supplier;
import com.example.merchmanager.service.SupplierService;
import org.springframework.beans.factory.annotation.Autowired;
import org.springframework.cache.annotation.CacheEvict;
import org.springframework.cache.annotation.Cacheable;
import org.springframework.stereotype.Service;

import java.util.List;

@Service
public class SupplierServiceImpl implements SupplierService {
    @Autowired
    SupplierDao supplierDao;
    @Override
    @Cacheable("supp")
    public List<Supplier> findAll() {
        System.out.println("SupplierServiceImpl.findAll");
        return supplierDao.findAll();
```

```
    }
    @Override
    @CacheEvict(cacheNames = "supp", allEntries = true)
    public Supplier save(Supplier entity) {
        System.out.println("SupplierServiceImpl.save");
        return supplierDao.save(entity);
    }
    @Override
    @CacheEvict(cacheNames = "supp", allEntries = true)
    public void deleteById(Integer id) {
        System.out.println("SupplierServiceImpl.deleteById");
        supplierDao.deleteById(id);
    }
}
```

5.5.3 运行程序测试缓存效果

在IDEA中启动项目，待项目启动成功后，在浏览器中访问http://127.0.0.1:8080/，页面如图5.9所示。

单击"供应商管理"链接，进入供应商管理页面，如图5.10所示。

图 5.9　商品管理系统主页

图 5.10　供应商列表

观察IDEA控制台，会发现进入方法的输出以及Hibernate的日志输出，如图5.11所示。

```
SupplierServiceImpl.findAll
Hibernate: select supplier0_.id as id1_1_, supplier0_.address as
 address2_1_, supplier0_.name as name3_1_ from supplier supplier0_
```

图 5.11　访问数据库的日志输出

这说明进入了findAll方法，并且访问了数据库。

然后刷新页面，页面显示正常，再观察IDEA控制，就会发现没有新的日志输出。更进一步，可以在Redis中查看。先通过keys命令搜索以"supp"开头的key，然后获取key相关的数据，具体如图5.12所示。

```
127.0.0.1:6379> keys supp*
supp::SimpleKey []
127.0.0.1:6379> get "supp::SimpleKey []"
**srjava.util.ArrayListx***Isjzexpwsr&com.example
.merchmanager.pojo.Supplier*2*@Laddresstljava.lang
/string;LidtLjava/lang/Integer;Lnameq~xpt文具街道123
号srjava.lang.Integer..*8Ivaluexrjava.lang.Number*
*
***xpt
文具A厂 x
```

图 5.12　在 Redis 中查看缓存数据

此时可以在页面上操作，比如增加供应商。在供应商页面单击"增加"按钮，并录入相关信息，最后单击"保存"按钮，具体如图5.13所示。

弹出成功提示框后，单击"确定"按钮，会发现列表中增加了一条数据，显示如图5.14所示。

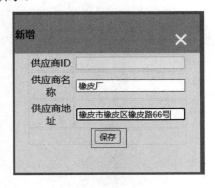

图 5.13　增加供应商　　　　　　　　　图 5.14　新供应商列表

注意，这里列表虽然新增了一条数据，但其实只是页面的操作，并没有访问列表接口，所以从后台来看，只是save(Supplier entity)方法被调用。因此，从逻辑上推断，缓存key为"supp::SimpleKey []"的缓存被删除。在此从Redis中执行命令查询以"supp"开头的缓存，结果和我们推测的相同，确实没有数据，如图5.15所示。

```
127.0.0.1:6379> keys supp*
127.0.0.1:6379>
```

图 5.15　Redis 中已删除缓存

此时，如果我们刷新页面，会发现控制台又会出现了方法中的日志输出，以及Hibernate日志输出。由此便可以证明我们的缓存增加和删除生效了。

第 6 章

Spring Boot整合Elasticsearch

Elasticsearch作为一款常见的搜索引擎，在项目中随时会用到。为方便在项目中访问Elasticsearch，Spring官方封装了对Elasticsearch的支持。通过Spring Data的Elasticsearch模块，我们可以方便地使用Elasticsearch。本节将介绍spring-data-elasticsearch项目和在Spring Boot项目中整合Elasticsearch。

本章主要涉及的知识点有：

- Elasticsearch的使用场景和技术。
- spring-data-elasticsearch中Operations和Repository相关技术的使用。
- 如何在Spring Boot项目中引入spring-data-elasticsearch。

6.1 Elasticsearch的使用场景和相关技术

搜索功能不仅在互联网项目中需要，在企业级项目中也需要。在通用型搜索引擎出现之前，通常实现搜索功能的方式是关系型数据库的模糊查询，但是使用模糊查询具有效率低、响应速度慢、不支持匹配度排序等缺陷。因此，在项目中引入搜索引擎就成了实现搜索功能的不二之选。

常用的搜索引擎除Elasticsearch之外还有Solr，它和Elasticsearch都是基于Lucene开发出来的。Lucene是Apache的开源搜索项目，这个项目的产出是一个搜索库，被称为Lucene Core。Lucene Core是Java实现的，提供了强大的索引和搜索功能，以及拼写检查、单击突出显示和高级分析/标记功能。

Apache Solr是Apache的一个独立的顶级项目，其内置了完整的Lucene包。自从Lucene和Solr整合之后，Solr和Lucene发布的版本都是一致的。

由于Lucene Core只是Java库，不能独立使用，因此平时在企业中使用最多的还是Elasticsearch或Solr。Elasticsearch和Solr都能实现搜索，但是也不完全相同。Solr有庞大的用户群，而且比较成熟，但是建立索引时会影响搜索效率，不适合用作实时搜索。而Elasticsearch

支持分布式、实时分发,支持建立索引和搜索同时进行。也正是由于Elasticsearch的这些特点,使得Elasticsearch的使用变得越来越流行。

6.2 spring-data-elasticsearch支持的Elasticsearch Client

要在项目中使用Elasticsearch,首先要连接到Elasticsearch。实现连接到Elasticsearch的模块被称作Elasticsearch Client。

6.2.1 Elasticsearch 的 Client

Elasticsearch官方提供了3个Client,具体如下:

```
org.elasticsearch.client.transport.TransportClient
org.elasticsearch.client.RestClient
org.elasticsearch.client.RestHighLevelClient
```

TransportClient位于Elasticsearch包下,是Elasticsearch官方早期支持的Elasticsearch Client,但是在Elasticsearch 7.x版本中已经标注为Deprecated,并且将在8.0版本中移除,所以建议不使用TransportClient作为Elasticsearch Client。

RestHighLevelClient是TransportClient的直接替代者,也是Elasticsearch官方推荐和默认的Client。

除了Elasticsearch官方提供的Client,spring-data-elasticsearch还支持响应式的客户端ReactiveElasticsearchClient。ReactiveElasticsearchClient是基于WebClient技术实现的Elasticsearch Client,依赖于Spring的响应式栈。响应式栈在本书中不会涉及。

6.2.2 创建 RestHighLevelClient

spring-data-elasticsearch提供了接口AbstractElasticsearchConfiguration,使用该接口可以非常方便地在容器中引入RestHighLevelClient。

AbstractElasticsearchConfiguration接口位于org.springframework.data.elasticsearch.config包下,可以按照如下方式来使用:

```
@Configuration
public class ESConfigutration extends AbstractElasticsearchConfiguration {
    @Override
    public RestHighLevelClient elasticsearchClient() {
        final ClientConfiguration clientConfiguration = ClientConfiguration.builder()
                .connectedTo("192.168.110.128:9200")
                .build();
        return RestClients.create(clientConfiguration).rest();
    }
}
```

如上面的代码，自定义一个配置类，继承AbstractElasticsearchConfiguration接口，并实现接口中定义的方法RestHighLevelClient elasticsearchClient()。

上面的ClientConfiguration用来配置Elasticsearch客户端的属性，比如可以配置代理、连接超时时长以及socket超时时长等，上面的代码示例中只配置了Elasticsearch服务的地址和端口号。

6.3　使用operations相关API操作Elasticsearch

spring-data-elasticsearch中定义了4个命名以Operations结尾的接口，用来操作Elasticsearch的不同对象，接口分别是IndexOperations、DocumentOperations、SearchOperations和ElasticsearchOperations。

6.3.1　4个Operations接口

IndexOperations和DocumentOperations接口从命名上可以看出，分别定义的是Index级别的接口和Document的接口，比如创建和删除索引（Index）以及根据id更新、查询entity。spring-data-elasticsearch提供了IndexOperations的默认实现，如图6.1所示，有DefaultIndexOperations和DefaultTransportIndexOperations两个实现类。这两个实现类使用的客户端不同，前者使用的是RestHighLevelClient，后者使用的是TransportClient，当然更推荐使用前者。

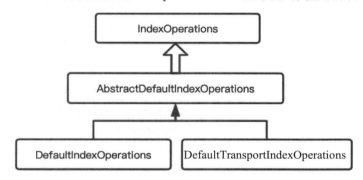

图 6.1　IndexOperations 接口的继承关系及其实现类

在spring-data-elasticsearch的4.2.3版本中，IndexOperations中定义了操作Index的方法，对应Elasticsearch官方文档的Index APIs部分，用于管理单个索引、索引设置、别名、映射和索引模板等。注意，表6.1中不包含已经被标记为过时的方法，以及不包含使用Elasticsearch Legacy API实现的方法。

表 6.1　接口 IndexOperations 方法列表

返 回 值	方法签名	方法描述
boolean	alias(AliasActions aliasActions)	执行给定的AliasActions对象
boolean	create()	创建索引
boolean	create(Map<String,Object> settings)	根据指定的settings设置创建索引

（续表）

返回值	方法签名	方法描述
boolean	create(Map<String,Object> settings, Document mapping)	根据指定的设置和索引映射创建索引
Document	createMapping()	创建当前IndexOperations实例绑定实体类的索引映射
Document	createMapping(Class<?> clazz)	创建指定类的索引映射
Settings	createSettings()	创建当前IndexOperations实例绑定实体类的设置对象
Settings	createSettings(Class<?> clazz)	创建指定类索引设置对象
boolean	createWithMapping()	使用当前对象的设置和映射信息创建索引
boolean	delete()	删除当前对象所绑定的索引
boolean	exists()	检查索引是否存在
Map<String,Set<AliasData>>	getAliases(String... aliasNames)	获取别名的信息
Map<String,Set<AliasData>>	getAliasesForIndex(String... indexNames)	获取索引的别名信息
IndexCoordinates	getIndexCoordinates()	IndexCoordinates中使用final关键字封装了索引名和type名称，不过在当前版本中type已经是常量值"_doc"
default List<IndexInformation>	getInformation()	获取由getIndexCoordinates()方法返回的索引的IndexInformation
Map<String,Object>	getMapping()	获取由实体类定义的索引信息的映射
Settings	getSettings()	获取索引的设置信息
Settings	getSettings(boolean includeDefaults)	获取索引的设置信息，参数用来指定是否包含所有的默认设置
default boolean	putMapping()	将当前实例关联实体类的映射信息写入索引中
default boolean	putMapping(Class<?> clazz)	创建索引并将映射信息写入
boolean	putMapping(Document mapping)	向索引中写入映射信息
void	refresh()	刷新索引

SearchOperation、DocumentOperations和ElasticsearchOperations的继承关系及其实现类的继承关系如图6.2所示。

从图6.2中可以看出，ElasticsearchOperations继承了SearchOperation和DocumentOperations，并且spring-data-elasticsearch为ElasticsearchOperations提供了两个实现类，即ElasticsearchRestTemplate和ElasticsearchTemplate，同上面的实现类一致，也是基于两种客户端的实现。

DocumentOperations接口中定义了Document级别的方法，对应Elasticsearch官方文档的Document APIs部分，包括对Document的增删改查等操作，具体方法列举在表6.2中。

第 6 章 Spring Boot 整合 Elasticsearch

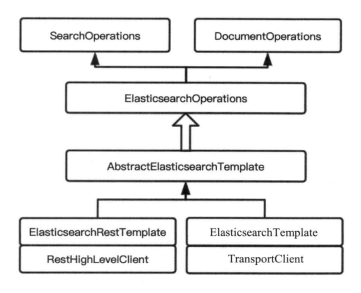

图 6.2　3 个 Operations 接口的继承关系及其实现类

表 6.2　接口 DocumentOperations 方法列表

返　回　值	方法签名	方法描述
List<IndexedObjectInformation>	bulkIndex(List<IndexQuery> queries, BulkOptions bulkOptions, Class<?> clazz)	批量检索所有对象
List<IndexedObjectInformation>	bulkIndex(List<IndexQuery> queries, BulkOptions bulkOptions, IndexCoordinates index)	批量检索所有对象
default List<IndexedObjectInformation>	bulkIndex(List<IndexQuery> queries, Class<?> clazz)	批量检索所有对象
default List<IndexedObjectInformation>	bulkIndex(List<IndexQuery> queries, IndexCoordinates index)	批量检索所有对象
void	bulkUpdate(List<UpdateQuery> queries, BulkOptions bulkOptions, IndexCoordinates index)	批量更新所有对象
void	bulkUpdate(List<UpdateQuery> queries, Class<?> clazz)	批量更新所有对象
default void	bulkUpdate(List<UpdateQuery> queries, IndexCoordinates index)	批量更新所有对象
String	delete(Object entity)	删除给定的对象。返回值是被删除的Document对象的id
String	delete(Object entity, IndexCoordinates index)	删除给定的对象，第二个参数指定删除对象所属的索引
ByQueryResponse	delete(Query query, Class<?> clazz)	删除所有匹配到的记录
ByQueryResponse	delete(Query query, Class<?> clazz, IndexCoordinates index)	删除所有匹配到的记录

（续表）

返回值	方法签名	方法描述
String	delete(String id, Class<?> entityType)	根据Document ID删除记录，并返回Document ID
String	delete(String id, IndexCoordinates index)	根据Document ID删除记录，并返回Document ID
boolean	exists(String id, Class<?> clazz)	检查指定的Document ID的实体是否存在
boolean	exists(String id, IndexCoordinates index)	检查指定的Document ID的实体是否存在
<T> T	get(String id, Class<T> clazz)	根据Document ID检索数据，并封装成实体类对象返回
<T> T	get(String id, Class<T> clazz, IndexCoordinates index)	根据Document ID检索数据，并封装成实体类对象返回
String	index(IndexQuery query, IndexCoordinates index)	创建或更新索引，返回被创建或更新的Document ID
<T> List<MultiGetItem<T>>	multiGet(Query query, Class<T> clazz)	执行multiGet操作
<T> List<MultiGetItem<T>>	multiGet(Query query, Class<T> clazz, IndexCoordinates index)	执行multiGet操作
<T> Iterable<T>	save(Iterable<T> entities)	将给定的实体对象保存到其类信息所关联的索引中
<T> Iterable<T>	save(Iterable<T> entities, IndexCoordinates index)	将给定的实体对象保存到指定的索引中
<T> Iterable<T>	save(T... entities)	将给定的实体对象保存到其类信息所关联的索引中
<T> T	save(T entity)	将给定的单个实体对象保存到其类信息所关联的索引中
<T> T	save(T entity, IndexCoordinates index)	将给定的单个实体对象保存到指定的索引中
UpdateResponse	update(UpdateQuery updateQuery, IndexCoordinates index)	根据查询部分更新文档
ByQueryResponse	updateByQuery(UpdateQuery updateQuery, IndexCoordinates index)	根据查询部分更新文档

接口SearchOperations中定义了搜索和聚合索引的相关操作，对应Elasticsearch官方文档的Search APIs部分，具体方法列举在表6.3中。

表 6.3　接口 SearchOperations 方法列表

返回值	方法签名	方法描述
long	count(Query query, Class<?> clazz)	获取符合查询条件的对象数量
long	count(Query query, Class<?> clazz, IndexCoordinates index)	获取符合查询条件的对象数量

（续表）

返 回 值	方法签名	方法描述
default long	count(Query query, IndexCoordinates index)	获取符合查询条件的对象数量
\<T\> List\<SearchHits\<T\>\>	multiSearch(List\<? extends Query\> queries, Class\<T\> clazz)	执行multi search查询
\<T\> List\<SearchHits\<T\>\>	multiSearch(List\<? extends Query\> queries, Class\<T\> clazz, IndexCoordinates index)	执行multi search查询
List\<SearchHits\<?\>\>	multiSearch(List\<? extends Query\> queries, List\<Class\<?\>\> classes)	执行multi search查询
List\<SearchHits\<?\>\>	multiSearch(List\<? extends Query\> queries, List\<Class\<?\>\> classes, IndexCoordinates index)	执行multi search查询
\<T\> SearchHits\<T\>	search(MoreLikeThisQuery query, Class\<T\> clazz)	类似于使用like关键字查询
\<T\> SearchHits\<T\>	search(MoreLikeThisQuery query, Class\<T\> clazz, IndexCoordinates index)	类似于使用like关键字查询
\<T\> SearchHits\<T\>	search(Query query, Class\<T\> clazz)	执行条件查询
\<T\> SearchHits\<T\>	search(Query query, Class\<T\> clazz, IndexCoordinates index)	执行条件查询
\<T\> SearchHitsIterator\<T\>	searchForStream(Query query, Class\<T\> clazz)	按照参数Query中封装的条件进行查询
\<T\> SearchHitsIterator\<T\>	searchForStream(Query query, Class\<T\> clazz, IndexCoordinates index)	按照参数Query中封装的条件进行查询
default \<T\> SearchHit\<T\>	searchOne(Query query, Class\<T\> clazz)	执行查询并返回结果中的第一条数据
default \<T\> SearchHit\<T\>	searchOne(Query query, Class\<T\> clazz, IndexCoordinates index)	执行查询并返回结果中的第一条数据
org.elasticsearch.action.search.SearchResponse	suggest(org.elasticsearch.search.suggest.SuggestBuilder suggestion, Class\<?\> clazz)	查询"查询建议"
org.elasticsearch.action.search.SearchResponse	suggest(org.elasticsearch.search.suggest.SuggestBuilder suggestion, IndexCoordinates index)	查询"查询建议"

最后，接口ElasticsearchOperations继承了DocumentOperations和SearchOperations，所以在接口DocumentOperations和SearchOperations中定义的方法都可以通过接口ElasticsearchOperations来使用。除了继承的方法外，接口ElasticsearchOperations中还定义了一些通用的辅助性的方法，具体列举在表6.4中。

表 6.4 接口 ElasticsearchOperations 方法列表

返 回 值	方法签名	方法描述
ClusterOperations	cluster()	返回一个ClusterOperations实例，该实例使用与ElasticsearchOperations实例相同的客户端通信设置

（续表）

返 回 值	方法签名	方法描述
ElasticsearchConverter	getElasticsearchConverter()	获取ElasticsearchConverter实例
String	getEntityRouting(Object entity)	获取由关联类型关系定义的实体的路由
IndexCoordinates	getIndexCoordinatesFor(Class<?> clazz)	获取索引的IndexCoordinates实例
IndexOperations	indexOps(Class<?> clazz)	获取绑定到给定类的IndexOperations实例
IndexOperations	indexOps(IndexCoordinates index)	获取绑定到给定IndexCoordinates实例的IndexOperations实例
default String	stringIdRepresentation(Object id)	获取参数id的String形式
ElasticsearchOperations	withRouting(RoutingResolver routingResolver)	使用参数指定的路由解析器和当前实例的配置对当前实例进行复制

6.3.2 搜索结果类型

Elasticsearch搜索API在返回搜索数据的同时也会返回搜索产生的额外信息，比如匹配到的总数量、排序字段值、高亮显示等，这些伴随着搜索的额外信息就被放置在spring-data-elasticsearch提供的搜索结果包装类中。本节将介绍spring-data-elasticsearch搜索结果的包装类。

由于部分类使用了和Elasticsearch官方提供的相同的类名，因此先对这些类所属的包说明一下，以下所提及的类如果没有特殊说明，默认都是包org.springframework.data.elasticsearch.core下的，而不是Elasticsearch官方的包org.elasticsearch.search下的同名类。

1. SearchHit<T>

搜索接口返回的数据实体都会使用SearchHit<T>类作为包装，用来放置数据实体相关的搜索信息，具体字段（get方法）信息见表6.5。

表 6.5 SearchHit<T>字段说明

字 段 名	类 型	描 述
id	String	文档id
score	Float	得分，搜索匹配度的一种体现，默认根据此字段降序排列
sortValues	Object[]	排序字段的值，只有当查询时指定了sort参数，此字段才有值
T	Content	实体类对象，范型指向实体类类型
highlightFields	Map<String, HighlightField>	高亮字段的值，也是当指定了高亮字段时才显示
innerHits	Map<String, SearchHits>	嵌套和关联查询时，内部命中的数据

2. SearchHits<T>

上面的SearchHit<T>是对单条数据的封装，而接口SearchHits<T>是对整体搜索结果的封装，其内部定义了获取SearchHit<T>列表的方法，以及获取一次搜索的总体数据的方法等，具体方法信息见表6.6。

表 6.6 SearchHits<T>接口方法说明

方法签名	返回值类型	描述
getTotalHits()	long	查询命中的总记录数量，该数值是否准确还需要getTotalHitsRelation()的返回值说明
getTotalHitsRelation()	TotalHitsRelation	说明getTotalHits()查询到的总数比实际可以查询到的数量是少还是相等
getSearchHits()	List<SearchHit<T>>	获取查询结果的SearchHits列表
getMaxScore()	float	获取符合查询条件的最高score，注意不是本次返回数据的最高score
getAggregations()	Aggregations	聚合结果

6.3.3 查询条件的封装

在接口SearchOperations中定义的方法，除了最后两个查询建议的方法外，其他方法中都使用了类型为org.springframework.data.elasticsearch.core.query.Query的参数。该类型为接口类型，spring-data-elasticsearch提供了3个实现类，分别是CriteriaQuery、StringQuery和NativeSearchQuery。下面将依次介绍这3个实现类。

1. CriteriaQuery

CriteriaQuery允许我们通过API调用的方式来定义查询条件，好处就是不需要用户理解Elasticsearch原生的查询语法。

CriteriaQuery有两个构造器，也是创建CriteriaQuery的两种方式。这两个构造器都需要类为org.springframework.data.elasticsearch.core.query.Criteria的封装查询条件，所以可以理解为CriteriaQuery只是Criteria的包装类，我们创建和封装查询条件主要通过Criteria来实现。

Criteria方法的命名仿照了SQL关键字，比如创建查询条件可以使用其静态方法where()。另外，对于多个条件组合，可以使用and()和or()。

下面通过示例说明CriteriaQuery类的使用方式。

【示例6.1】 查询出版时间为给定年份的图书

假定publishYear字段为图书出版年份，那么查询出版年份为2021年的图书的查询条件封装的代码如下：

```
Criteria criteria = Criteria.where("publishYear").is(2021);
Query query = new CriteriaQuery(criteria);
```

【示例6.2】 查询出版时间在2015~2019年，并且类别为科学技术的图书

使用静态方法and()创建第二个条件并与第一个条件关联起来，具体代码如下：

```
Criteria criteria = Criteria.where("publishYear").between(2015,2019).
and("category").is("科学技术");
CriteriaQuery criteriaQuery = new CriteriaQuery(criteria);
```

【示例6.3】 查询出版时间在2015~2019年，并且类别为科学技术或历史人文的图书

这里用到了或和与的组合，使用subCriteria来实现，具体代码如下：

```
Criteria criteria = Criteria.where("publishYear").between(2015, 2019).
subCriteria(Criteria.where("category").is("科学技术").or("category").is("历史人文"));
CriteriaQuery criteriaQuery = new CriteriaQuery(criteria);
```

2. StringQuery

StringQuery以Elasticsearch可以理解的JSON格式封装查询条件，因此比较适合熟悉Elasticsearch查询语法的用户。

这里的JSON格式可以对应Elasticsearch的Query DSL（Domain Specific Language）语法，具体可以参考Elasticsearch官方文档的Query DSL部分。

【示例6.4】 使用StringQuery查询出版时间在2015~2019年，并且类别为历史人文的图书使用到了Boolean查询来组合两个查询条件，具体代码如下：

```
Query query = new StringQuery("{\n" +
    "  \"bool\": {\n" +
    "    \"must\": [\n" +
    "      {\n" +
    "        \"match\": {\n" +
    "          \"category\": \"历史人文\"\n" +
    "        }\n" +
    "      },\n" +
    "      {\n" +
    "        \"range\": {\n" +
    "          \"publishYear\": {\n" +
    "            \"gte\": 2015,\n" +
    "            \"lte\": 2019\n" +
    "          }\n" +
    "        }\n" +
    "      }\n" +
    "    ]\n" +
    "  }\n" +
    "}");
SearchHits<Book> searchHits = operations.search(query, Book.class);
```

从上面的例子可以看出，使用StringQuery虽然创建对象简单，但是可读性差，并且需要熟悉Elasticsearch的DSL语法。

3. NativeSearchQuery

NativeSearchQuery使用实现比较复杂的查询，比如聚合操作等。由于其可以和Elasticsearch官方API结合使用，因此命名为Native。

虽然从功能上讲NativeSearchQuery比CriteriaQuery强大，但是由于其使用既需要熟悉Elasticsearch官方API，又需要学习NativeSearchQuery的API，学习成本相比前两种要更高一些，并且调试起来也不比DSL容易，所以也没有比较明显的优势。

在工作中，简单的查询使用CriteriaQuery，复杂的查询先在Kibana中使用DSL调试好查询语句，然后直接复制到代码中创建StringQuery来构建查询，这样或许效率更高一些。

6.4 Repository的使用

spring-data-elasticsearch中Repository的概念和功能与spring-data的Repository基本一致，是对spring-data的Repository概念的实现和拓展。Repository这一概念在spring-data中包括两大功能：

- 对实体类（Entities）的管理。
- 基于实体类的增删改查操作的定义和使用。

第二个功能基于第一个功能来完成，本节将介绍这两个功能及其使用。

6.4.1 使用注解管理索引实体类

类似于sping-data-jpa中使用@Table来关联表和实体类，在spring-data-elasticsearch中使用@Document来关联索引和实体类。注解@Document标注的实体类会被映射到Elasticsearch的索引中，其属性indexName用来标注关联的索引名称。注解@Document的另一个属性createIndex用来标识是否在项目启动时，当Elasticsearch中的索引不存在时创建索引。

当使用注解@Document标注实体类之后，通常必不可少的是指定索引的主键。使用注解@org.springframework.data.annotation.Id来标识索引的主键，这里需要与JPA使用的主键标识注解@javax.persistence.Id相区分，所以也就意味着当spring-data-jpa和spring-data-elasticsearch共用一个实体类时，实体类的id上需要同时标注两个注解，即@org.springframework.data.annotation.Id和@javax.persistence.Id。

被注解@Document标注的实体类，其字段默认以其名称被映射到索引的field。如果不想某个字段被映射，可以使用注解@Transient标注其字段，以忽略被标注的字段。

另外，还可以使用注解@Field来自定义字段级别的映射信息。比如，可以使用name属性定义Elasticsearch索引的field名称。还可以使用type属性指定filed的数据类型，可选有Text、Keyword、Long、Integer、Short、Byte、Double、Float、Date、Boolean、Binary、Object、Ip和Flattened等。另外，还可以使用属性analyzer、searchAnalyzer和normalizer等来指定分析器和规范化器。

Type Hints是spring-data-elasticsearch向文档中添加的名称为"_class"的一个字段，其值默认为原始Java类的全限定名，这个字段用来标记文档所对应的Java类，以供解析数据时由反序列化器来使用。比如，实体类Book代码如下：

```
package com.demo;
import org.springframework.data.annotation.Id;
import org.springframework.data.elasticsearch.annotations.Document;
@Document(indexName = "book")
public class Book{
    @Id
    private Long id;
    private String title;
```

```
    private String content;

    public Book(){}

    public Book(Long id, String title, String content) {
        this.id = id;
        this.title = title;
        this.content = content;
    }
    public Long getId() {
        return id;
    }
    public String getTitle() {
        return title;
    }
    public String getContent() {
        return content;
    }
}
```

默认对应的索引文档的JSON文本如下：

```
{
  "_class" : "com.demo.Book",
  "id" : 1,
  "title" : "书名...",
  "content" : "图书内容简介..."
}
```

可以通过注解@TypeAlias自定义Type Hints的内容，比如在Book类上使用@TypeAlias注解如下：

```
...
@TypeAlias("book")
@Document(indexName = "book")
public class Book{
    @Id
    private Long id;
    private String title;
    private String content;
    ...
}
```

然后新生成的文档内容如下：

```
{
  "_class" : "book",
  "id" : 1,
  "title" : "书名...",
  "content" : "图书内容简介..."
}
```

6.4.2 查询方法的定义

本节将介绍如何在Repository子接口中通过方法名来定义查询。

使用方法名定义查询的格式基本与spring-data中的Repository的格式是一致的，首先需要定义接口，并继承Repository或Repository的子接口，这里仍以Book实体类为例，代码如下：

```
interface BookRepository extends Repository<Book, Long> {}
```

1. 方法名的定义

在接口中定义方法，方法名的第一部分是subject keywords，也就是说需要定义方法要做什么事，可选有findBy、findFirstBy、countBy、existsBy、deleteBy等。

方法名的第二部分是查询关键字，spring-data-elasticsearch中支持的查询关键字列举在表6.7中。

表 6.7　Repository 查询关键字

关 键 字	作用描述
And	用于关联两个关键字，转换为使用must组合两个查询条件
Or	用于关联两个关键字，转换为使用should组合两个查询条件
Is	使用query_string，注意不是相等判断
Not	对条件否定，会转换为must_not条件
Between	查询在两者之间，转换为range条件
LessThan	小于指定值
LessThanEqual	小于或等于指定值
GreaterThan	大于指定值
GreaterThanEqual	大于或等于指定值
Before	小于指定值
After	大于指定值
Like	使用通配符查询，会自动在参数末尾拼接上通配符"*"
StartingWith	使用通配符查询，会自动在参数末尾拼接上通配符"*"
EndingWith	使用通配符查询，会自动在参数开头拼接上通配符"*"
Contains/Containing	使用通配符查询，会自动在参数开头和结尾拼接上通配符"*"
In	对标注为keyword的字段使用terms查询，非标明字段使用query_string查询
NotIn	相当于在In的基础上使用了must_not
True	查询字段内容为true的数据
False	查询字段内容为false的数据
OrderBy	指定排序字段和排序方向，对应条件sort

2. 返回值类型的选择

可以通过返回值类型来选择是接收分页还是只接收数据，这里支持的返回值类型可以是 Page<T>、List<T>、Stream<T>、SearchHits<T>、List<SearchHit<T>>、Stream<SearchHit<T>> 和 SearchPage<T>。

【示例6.5】 定义查询方法查询出版时间在2015~2019年，并且类别为历史人文的图书

方法名第一部分使用findBy。第二部分由两个条件组成，使用到Is关键字和Between关键字，并且使用And组合。返回值可以使用SearchHits<T>，具体代码如下：

```
SearchHits<Book> findByCategoryIsAndPublishYearBetween("历史人文", 2015, 2019);
```

6.4.3 使用@Query注解定义查询

在Repository的子接口中，可以使用@Query标注在方法上定义查询。@Query的value属性指定内容与创建StringQuery时传递的字符串基本一致，但使用@Query的好处是不需要我们拼接参数，可以使用问号拼接数字来从形参中取值，以及可以通过Pageable类型参数作为形参来实现分页，无须在字符串中拼接。

【示例6.6】 使用@Query定义查询出版时间在2015~2019年，并且类别为历史人文的图书

使用@Query方法名不受限制，方法的定义代码如下：

```
@Query("{\n" +
    "  \"bool\": {\n" +
    "    \"must\": [\n" +
    "      {\n" +
    "        \"match\": {\n" +
    "          \"category\": \"?0\"\n" +
    "        }\n" +
    "      },\n" +
    "      {\n" +
    "        \"range\": {\n" +
    "          \"publishYear\": {\n" +
    "            \"gte\": ?1,\n" +
    "            \"lte\": ?2\n" +
    "          }\n" +
    "        }\n" +
    "      }\n" +
    "    ]\n" +
    "  }\n" +
    "}")
Page<Book> findByCategoryAndPublic(String category, int from, int to, Pageable pageable);
```

在调用查询的类中，我们需要先注入接口，比如使用注解注入，代码如下：

```
@Autowired
BookRepository bookRepository;
```

然后通过实例调用方法,并且指定按照publishYear升序排序,代码如下:

```
Sort sort = Sort.by("publishYear").ascending();
Pageable pageable = PageRequest.of(0, 5, sort);
bookRepository.findByCategoryAndPublic("科学技术", 2015, 2019, pageable);
```

6.5 在Spring Boot中配置spring-data-elasticsearch

在pom.xml中配置spring-data-elasticsearch提供的Starter,以此来启用spring-data-elasticsearch,配置代码如下:

```xml
<dependency>
    <groupId>org.springframework.boot</groupId>
    <artifactId>spring-boot-starter-data-elasticsearch</artifactId>
</dependency>
```

接下来通过配置来创建Elasticsearch客户端,推荐使用RestHighLevelClient,具体代码如下:

```java
@Configuration
public class ESConfigutration extends AbstractElasticsearchConfiguration {
    private String hostAndPort = "127.0.0.1:9200";
    @Override
    public RestHighLevelClient elasticsearchClient() {
        final ClientConfiguration clientConfiguration =
ClientConfiguration.builder()
                .connectedTo(hostAndPort)
                .build();
        return RestClients.create(clientConfiguration).rest();
    }
}
```

通过上面的配置已经可以在项目中使用RestHighLevelClient、ElasticsearchRestTemplate和Operations接口等。但是如果需要使用Repository的功能,还需要通过注解@EnableElasticsearchRepositories指定Repository子接口所在的包,比如如下配置类代码:

```java
@EnableElasticsearchRepositories(basePackages = "com.demo.learn.repository")
@Configuration
public class ESConfigutration extends AbstractElasticsearchConfiguration {
```

第 7 章

Spring Boot的日志管理

日志通常是生产环境必不可少的功能，通过日志我们可以监控项目的运行，记录运行中的数据，尤其是在项目出现异常，查找导致异常的问题时，日志更是离不开的工具。Spring Boot默认开启了对日志的支持，本章将介绍Spring Boot日志管理的相关内容。

本章主要涉及的知识点有：

- ⌘ 了解常用的日志框架及常用术语。
- ⌘ 了解Spring Boot的日志配置项。
- ⌘ 如何配置控制台日志输出。
- ⌘ 如何配置输出到日志文件。

7.1 常用的日志框架

Spring Boot对常用的日志框架都有支持，也都可以方便地配置和使用。如果不了解这些常用的日志框架，可能对Spring Boot中的日志配置懵懵懂懂，所以有必要在学习Spring Boot的日志管理之前，先对常用的日志框架做一定的了解。本节将对一些日志框架和相关术语做基本的介绍。

7.1.1 日志实现

Java日志框架有许多，使用比较早且使用最广泛的毫无疑问的是Log4j。不过，日志服务项目管理委员会已经在2015年8月5日宣布结束对Log4j的更新，并且推出了Log4j的升级版本，即Log4j 2。

在Log4j被广泛使用之后，从JDK 1.4开始，Java官方也推出了内置于JDK的日志API，即Java Util Logging（JUL）。Java Util Logging位于java.util.logging包下，提供了用于记录日志的一系列接口和实现类。Java Util Logging的使用也比较简单，而且由于内置于JDK，因此不用单独添加进来。不过由于Java Util Logging提供的时间比Log4j晚，而且功能上与Log4j相同，因此Java Util Logging在项目中被使用的流行程度一直不及Log4j。

Log4j的出现比较早，由于其性能比较低，所以目前项目中很少会主动选择使用Log4j，取而代之的是Logback。Logback也是一款日志框架，而且是当前最流行的日志框架。Logback的作者也是Log4j的创始人，Logback出现的时间比Log4j晚，所以可以认为Logback是Log4j的升级版本，而且Logback的性能要比Log4j好一些。

Logback相比于Log4j，性能大概高了一倍左右。Logback作为Log4j的替代者，被广泛地应用于项目中。不过Logback的性能并非日志框架的极限，还有很大的潜力可以挖掘。Apache于2015年发布了Log4j的升级版本，也就是前面提到过的Log4j 2。

Log4j 2是对已经存在的日志框架的一次重大升级。Log4j 2在设计和实现上借鉴了Logback，对Logback的一些问题做了修复，相较于Logback和Log4j，性能有了十分显著的提高。Log4j 2在性能方面的提高非常显眼，根据Log4j 2官方提供的数据，Log4j 2相较于Logback，日志输出效率的提高超过了10倍。并且Log4j 2基于零垃圾的对象复用实现，最大限度地避免JVM GC的出现，极大程度地提高了程序的执行效率。

由于极高的性能，Log4j 2很可能成为继Log4j和Logback之后被广泛应用的日志框架。

7.1.2 日志门面

所谓日志门面，就是一套日志API，类似于JDBC是操作关系型数据库的抽象。根据面向接口编程的思想，日志门面框架只定义日志接口，具体实现由日志框架来提供。使用日志门面技术，尤其是在编写框架（或库）和使用框架时变得非常方便。比如，在编写框架时，只使用日志门面的API即可，而不用指定具体的日志实现，这样框架在被引入项目中时，使用项目的日志框架即可实现日志功能，而不会出现多个框架需要以多个日志实现的情况，提高了项目开发效率。

目前流行的日志门面有3个，分别是SLF4J、Commons-logging（JCL）和Log4j 2。Commons-logging和Log4j 2是Apache推出的，SLF4J是Log4j和Logback的作者推出的。

Commons-logging也有人称作JCL，是Apache推出的日志门面技术。Commons-logging最早发布于2002年，也是三者中最早的日志门面技术，对Log4j和Java Util Logging这两种日志实现框架做了抽象。Commons-logging虽然也支持用户自定义其他日志框架的适配器，不过相比于原生支持Logback的SLF4j，其流行度还是要差一些，并且Commons-logging最近一次发布新版本还是在2014年，所以可以预料到Commons-logging无论是在当前还是在未来，其流行度都很难超过SLF4J。不过，Spring Boot内部使用的日志门面框架是Commons-logging。

SLF4J和Log4j、Logback同一个作者，也是目前最流行的日志门面框架。SLF4J最早发布于2005年，至今仍然在更新中，并且在2021年7月2日发布了2.0.0版本。SLF4J支持多种日志框架，比如Java Util Logging、Logback和Log4j等，也支持Log4j 2。SLF4J的流行得益于对Logback的原生支持，从今后的发展趋势来看，或许SLF4J和Log4j 2的日志实现是不错的搭配。

Log4j 2在设计上借鉴了SLF4J，分成了接口和实现两个模块。Log4j API是Log4j 2的接口定义，可以独立于具体实现作为日志门面使用。Log4j API支持Log4j 1.2、SLF4J、Commons-logging和Java Util Logging（JUL）等。Log4j API还有很好的兼容性，任何兼容SLF4J的日志框架都可以使用log4j-to-slf4j适配器来对接到Log4j API。

与SLF4J相比，Log4j API官方梳理了Log4j API具有的优势：

- Log4j API 支持多种消息格式，不仅仅是字符串。
- 支持拉姆达表达式。
- 丰富的日志方法。
- 支持使用 java.text.MessageFormat 语法和 printf 样式的消息的事件。
- 支持 LogManager.shutdown()方法。
- 完全支持其他构造，如标记、日志级别和 ThreadContext（也就是MDC）。

所以Log4j API作为门面技术，对SLF4J的替代性还是非常强的。在项目中，尤其是使用了Log4j 2作为实现时，使用Log4j API作为日志门面是一种不错的选择。

7.2 Spring Boot支持的日志配置

Spring Boot内部日志门面使用的是Commons-logging的Spring封装版本，即spring-jcl。Spring Boot内部并没有指定日志实现框架，具体日志实现对日志实现框架开放。本节将介绍Spring Boot对日志框架的支持。

7.2.1 Spring Boot 默认支持的日志框架

前面讲到Spring Boot的日志门面使用了spring-jcl。spring-jcl是对Commons-logging的封装，所以毫无疑问，我们使用Spring Boot构建项目时也可以用到spring-jcl。

使用Spring Boot构建项目，直接或间接会引用到spring-boot-starter，而spring-boot-starter依赖spring-core，进而spring-core依赖spring-jcl，这是对spring-jcl的依赖。除了依赖spring-jcl外，这里还有众多的日志框架。spring-boot-starter依赖spring-boot-starter-logging，从命名上可以看出，spring-boot-starter-logging是管理日志依赖的主要库。在spring-boot-starter-logging中定义了对Log4j 2、Logback和jul-to-slf4j的依赖，所以在使用Spring Boot构建项目时，在代码中可以直接使用Commons-logging、Log4j 2、Logback和Java Util Logging作为日志实现框架。

Spring Boot 2.5.3版本中默认引入的日志框架如图7.1所示。

从图7.1中可以看到，日志门面虽然引入了log4j-api，但是并没有引入log4j 2的日志实现。所以，如果需要使用log4j 2作为日志框架，还需要手动引入log4j-core的依赖。

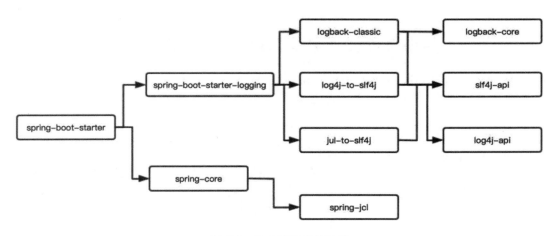

图 7.1　日志框架依赖关系

7.2.2　自定义日志配置

Spring Boot以及Spring容器自身也使用了日志框架，所以在项目中，日志框架的初始化在Spring容器初始化之前，无法在配置类中使用诸如@PropertySources的注解，可行的方案是直接使用System Properties或者将配置转换为System Properties，以此来配置日志框架。

1．通过系统属性指定日志实现框架

Spring Boot支持使用System Property来指定日志实现框架，这个系统属性为org.springframework.boot.logging.LoggingSystem，通过为这个系统属性指定LoggingSystem实现类的完全限定名来指定日志实现框架，或者将该值指定为none来禁用日志框架。具体步骤可以参考下面示例来学习。

【示例7.1】　通过系统参数设置日志实现框架使用Log4j 2

我们要启动Log4j 2的日志实现框架，但是默认条件下，Spring Boot没有引入log4j-core的包，所以需要手动引入依赖，在pom.xml中引入，具体配置代码如下：

```xml
<dependency>
    <groupId>org.apache.logging.log4j</groupId>
    <artifactId>log4j-core</artifactId>
    <version>2.14.1</version>
</dependency>
```

在程序启动时，设置系统参数指定日志实现框架的LoggingSystem实现类，来强制变更日志实现框架。这里要更换为Log4j2，所以应该指定的是Log4j2的LoggingSystem实现类，即Log4J2LoggingSystem。

设置System Property的可选方案是在开发工具中设置VM参数，如果是IDEA，可以通过Edit Configurations...来修改，如图7.2所示。

然后在如图7.3所示的位置添加配置"-Dorg.springframework.boot.logging.LoggingSystem=org.springframework.boot.logging.log4j2.Log4J2LoggingSystem"，最后单击OK按钮保存配置并关闭窗口。

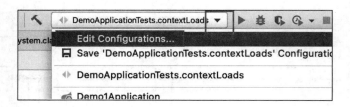

图 7.2　Edit Configurations

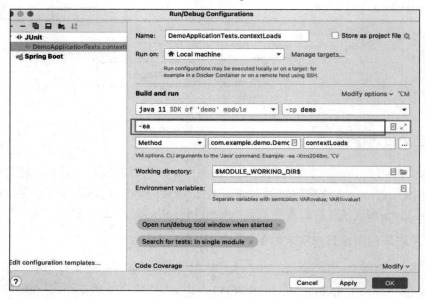

图 7.3　添加 JVM 参数

创建测试类，代码如下：

```
package com.example.demo;

import org.apache.commons.logging.Log;
import org.apache.commons.logging.LogFactory;
import org.junit.jupiter.api.Test;
import org.springframework.boot.test.context.SpringBootTest;

@SpringBootTest
class DemoApplicationTests {
    private static final Log logger = LogFactory.getLog(DemoApplicationTests.class);
    @Test
    void contextLoads() {
        logger.trace("测试日志输出 trace");
        logger.debug("测试日志输出 debug");
        logger.info("测试日志输出 info");
        logger.warn("测试日志输出 warn");
        logger.error("测试日志输出 error");
        System.out.println(logger.getClass().toString());
    }
}
```

此时，执行代码测试方法contextLoad，会发现控制台报错如下：

```
Caused by: java.lang.ClassCastException: class org.apache.logging.slf4j.
SLF4JLoggerContext cannot be cast to class org.apache.logging.log4j.core.
LoggerContext (org.apache.logging.slf4j.SLF4JLoggerContext and org.apache.
logging.log4j.core.LoggerContext are in unnamed module of loader 'app')
```

这是由于SLF4J和Log4j 2发生了冲突，因此我们需要先解决冲突。解决方案是在spring-boot-starter的配置中排除掉对SLF4J的依赖。排除SLF4J依赖在pom.xml中进行配置，配置代码如下：

```xml
<dependency>
    <groupId>org.springframework.boot</groupId>
    <artifactId>spring-boot-starter</artifactId>
    <exclusions>
        <exclusion>
            <groupId>org.slf4j</groupId>
            <artifactId>slf4j-api</artifactId>
        </exclusion>
        <exclusion>
            <groupId>org.apache.logging.log4j</groupId>
            <artifactId>log4j-to-slf4j</artifactId>
        </exclusion>
    </exclusions>
</dependency>
```

因为这里使用的仅仅是spring-boot-starter，所以在配置文件中已经有了spring-boot-starter的配置，如果是Web项目，那么可以在spring-boot-starter-web下进行排除的配置。

在配置完成之后，运行测试方法，发现控制台输出日志并打印logger对象的类型，此时已经是Log4jLog类型。控制台输出如下：

```
2021-08-09 20:04:24.494  INFO 2722 --- [           main] 
c.e.d.DemoApplicationTests               : 测试日志输出 info
2021-08-09 20:04:24.495  WARN 2722 --- [           main] 
c.e.d.DemoApplicationTests               : 测试日志输出 warn
2021-08-09 20:04:24.496 ERROR 2722 --- [           main] 
c.e.d.DemoApplicationTests               : 测试日志输出 error
class org.apache.commons.logging.LogAdapter$Log4jLog
```

2. 通过 Starter 修改日志实现类

如果项目中需要使用Log4j 2代替Spring Boot默认的Logback，更为便捷的方式是使用Log4j 2的Spring Boot Starter。

首先需要排除掉默认依赖的日志Starter，即排除掉spring-boot-starter-logging。在pom.xml中进行配置，配置代码如下：

```xml
<dependency>
    <groupId>org.springframework.boot</groupId>
    <artifactId>spring-boot-starter</artifactId>
    <exclusions>
```

```xml
        <exclusion>
            <groupId>org.springframework.boot</groupId>
            <artifactId>spring-boot-starter-logging</artifactId>
        </exclusion>
    </exclusions>
</dependency>
```

与上面的示例类似，如果是Web项目，也可以在spring-boot-starter-web中配置排除。

然后引入对应Log4j 2的Starter，需要在pom.xml中添加依赖，配置代码如下：

```xml
<dependency>
    <groupId>org.springframework.boot</groupId>
    <artifactId>spring-boot-starter-log4j2</artifactId>
</dependency>
```

【示例7.2】 测试spring-boot-starter-log4j2配置的效果

创建测试类，同时创建Commons-logging和SLF4J的Logger对象，打印Logger对象的类信息。首先创建测试类，代码如下：

```java
package com.example.demo;

import org.apache.commons.logging.Log;
import org.apache.commons.logging.LogFactory;
import org.junit.jupiter.api.Test;
import org.slf4j.Logger;
import org.slf4j.LoggerFactory;
import org.springframework.boot.test.context.SpringBootTest;

@SpringBootTest
class DemoApplicationTests {
    private static final Log logger = LogFactory.getLog(DemoApplicationTests.class);
    private static final Logger loggerOfSLF4J = LoggerFactory.getLogger(DemoApplicationTests.class);
    @Test
    void contextLoads() {
        logger.trace("测试日志输出 trace");
        logger.debug("测试日志输出 debug");
        logger.info("测试日志输出 info");
        logger.warn("测试日志输出 warn");
        logger.error("测试日志输出 error");
        System.out.println(logger.getClass().toString());
        loggerOfSLF4J.trace("测试日志输出 trace");
        loggerOfSLF4J.debug("测试日志输出 debug");
        loggerOfSLF4J.info("测试日志输出 info");
        loggerOfSLF4J.warn("测试日志输出 warn");
        loggerOfSLF4J.error("测试日志输出 error");
        System.out.println(loggerOfSLF4J.getClass().toString());
    }
}
```

根据上面的排除和依赖配置修改pom.xml文件，执行测试类代码，控制台输出结果如下：

```
    2021-08-09 20:24:37.207  INFO 2784 --- [           main]
c.e.d.DemoApplicationTests               : 测试日志输出 info
    2021-08-09 20:24:37.208  WARN 2784 --- [           main]
c.e.d.DemoApplicationTests               : 测试日志输出 warn
    2021-08-09 20:24:37.209 ERROR 2784 --- [           main]
c.e.d.DemoApplicationTests               : 测试日志输出 error
    class org.apache.commons.logging.LogAdapter$Log4jLog
    2021-08-09 20:24:37.210  INFO 2784 --- [           main]
c.e.d.DemoApplicationTests               : 测试日志输出 info
    2021-08-09 20:24:37.210  WARN 2784 --- [           main]
c.e.d.DemoApplicationTests               : 测试日志输出 warn
    2021-08-09 20:24:37.211 ERROR 2784 --- [           main]
c.e.d.DemoApplicationTests               : 测试日志输出 error
    class org.apache.logging.slf4j.Log4jLogger
```

两个日志对象输出格式相同，并且都为Log4j的Logger对象，所以配置有效。

综上上面的测试可以看出，在我们的项目中不需要Logback时，使用Log4j Starter的方式更简单高效。

7.2.3 日志框架的配置文件

Spring Boot支持日志框架使用独立的配置文件，默认会从类路径的根目录下读取确定命名的配置文件，具体的默认名称在下面会提到。配置文件也可以通过Spring Environment的logging.config来制定，这个将在7.2.4节提及。

不同日志实现的框架，默认加载的配置文件名称不同，即使是同一种框架，也支持不同名称的配置文件，具体参见表7.1。

表 7.1 Spring Boot 默认加载的日志配置文件名称

日志框架	描述支持的文件名称
Log4j 2	log4j2-spring.xml或log4j2.xml
Logback	logback-spring.xml、logback-spring.groovy、logback.xml或logback.groovy
Java Util Logging	logging.properties

从表7.1可以看出，对于Log4j 2和Logback，同一类型的配置文件都有两个命名，一个是带"-spring"的，另一个是不带的。两者不完全相同，带"-spring"的配置文件会交由Spring处理。比如，在带"-spring"的配置文件中，可以使用"<springProfile >"标签来定义所属不同环境的配置。由于其更丰富的功能，还是推荐使用带"-spring"的文件命名。

7.2.4 配置项汇总

Spring Boot默认配置支持Java Util Logging、Log4J2和Logback，并且默认的日志实现为Logback。

前面提到过日志框架的配置通过System Properties来完成，对比比较基础的配置，Spring Boot支持将Spring Environment转换成System Properties，从而当只是需要进行一些基础的配置时，我们也就没必要使用单独的日志配置文件。

不妨先概览Spring Boot支持的所有日志配置项，然后就常用的日志配置项进行说明。Spring Boot支持的所有日志配置项列举在表7.2中。

表 7.2 Spring Boot 日志配置项

配 置 项	SystemProperty	默 认 值	描 述
logging.exception-conversion-word	LOG_EXCEPTION_CONVERSION_WORD	%wEx	记录异常时使用的转换词
logging.file.name	LOG_FILE		输出的日志文件名
logging.file.path	LOG_PATH		输出的日志文件的目录
logging.pattern.console	CONSOLE_LOG_PATTERN	%clr(%d{${LOG_DATEFORMAT_PATTERN: yyyy-MM-ddHH:mm:ss.SSS}}){blue} %clr(${LOG_LEVEL_PATTERN:%5p}) %clr(${PID:-}){magenta} %clr(---){faint}%clr([%15.15t]){faint}%clr(%-40.40logger{39}){cyan}%clr(:){faint} %m%n${LOG_EXCEPTION_CONVERSION_WORD:%wEx}	输出在控制台的日志内容和格式，仅被Logback支持
logging.pattern.dateformat	LOG_DATEFORMAT_PATTERN	yyyy-MM-dd HH:mm:ss.SSS	日志日期格式，仅被Logback支持
logging.charset.console	CONSOLE_LOG_CHARSET		控制台输出的字符集
logging.pattern.file	FILE_LOG_PATTERN	%d{${LOG_DATEFORMAT_PATTERN:-yyyy-MM-ddHH:mm:ss.SSS}} ${LOG_LEVEL_PATTERN:-%5p} ${PID:- } --- [%t] %-40.40logger{39} : %m%n${LOG_EXCEPTION_CONVERSION_WORD:-%wEx}	输出在日志文件的日志内容和格式，仅被Logback支持
logging.charset.file	FILE_LOG_CHARSET		日志文件中输出的字符集
logging.pattern.level	LOG_LEVEL_PATTERN	%5p	指定某一级别的日志的格式和内容
logging.config			指定日志框架配置文件的位置
logging.group.*			用来设置属于一个组的多个Logger对象
logging.level.*			配置Log日志的级别

（续表）

配　置　项	SystemProperty	默　认　值	描　述
logging.logback.rollingpolicy.clean-history-on-start		false	是否在项目启动时清除已经打包的日志文件
logging.logback.rollingpolicy.file-name-pattern		${LOG_FILE}.%d{yyyyMM-dd}.%i.gz	滚动的日志文件命名
logging.logback.rollingpolicy.max-file-size		10MB	最大的日志文件大小
logging.logback.rollingpolicy.max-history		7	日志文件保存的最大天数
logging.logback.rollingpolicy.total-size-cap		0B	保存的日志总大小

上面的配置中，带有logback的都是仅支持Logback，其余的配置中有一部分也是只支持Logback，已经在列表中标明。

7.2.5　日志级别

日志级别通过"logging.level.*"来指定，其中"*"为Logger的name，也就是包名，即定义Logger时指定的class的包，可以参见下面的示例。

【示例7.3】 设置测试类中的日志级别以及Spring框架的日志级别为TRACE

测试类中定义Logger的代码如下：

```
package com.example.demo;

import org.apache.commons.logging.Log;
import org.apache.commons.logging.LogFactory;
import org.springframework.boot.test.context.SpringBootTest;

@SpringBootTest
class DemoApplicationTests {
    private static final Log logger = LogFactory.getLog(DemoApplicationTests.class);
    ...
}
```

所以，Logger的name可以是com.example.demo。另外，Spring框架所在的包可以指定为org.springframework，综合起来，配置代码如下：

```
logging:
  level:
    com.example.demo: "trace"
    com.springframework: "trace"
```

然后运行测试类，可以看到Spring日志和自定义的日志都打印到了TRACE级别，具体如图7.4所示。

图 7.4　日志打印到 trace 级别

上面的o.s.b.f.s.其实是org.springframework.beans.factory.support的首字母，其原因是默认日志格式指定了日志包名的长度。

级别的可选值有TRACE、DEBUG、INFO、WARN、ERROR、FATAL和OFF。这些级别虽然不一定会被日志框架全部支持，但是Spring Boot做了转换，我们可以放心使用。比如Logback没有FATAL，这个级别被映射到了ERROR。

7.2.6　日志格式和内容

关于日志格式和内容，从上面的配置项列表中可以看到有两个配置项可以用来修改日志格式，即logging.pattern.console和logging.pattern.file，分别用来修改输出到控制台的日志格式和输出到日志文件的日志格式。

Spring Boot定义了日志的默认格式和内容，具体如下：

日期和时间，默认精确到毫秒；
日志级别，可能是 ERROR、WARN、INFO、DEBUG 或 TRACE；
Process ID；
分隔符，默认为"---"；
线程名，包裹在中括号内；
logger 名称，包名和类型，包名过长时缩写；
日志信息。

比如日志输出代码如下：

`logger.info("测试日志输出 info");`

控制台的打印内容如图7.5所示。

图 7.5　默认日志输出效果

除了设置完整的日志格式外，还有两个配置项可以单独设置日志的一部分格式，即logging.pattern.level和logging.pattern.dateformat，它们都是在默认配置的基础上做微调。

logging.pattern.level用来配置日志级别的显示，默认值是%5p，其含义为日志级别右对齐，左边补充空格。常用的值还有%-5p和%p，分别表示左对齐，右边补充空格，以及不对齐。

【示例7.4】 观察logging.pattern.level设置不同值的效果

先观察默认值%5p的打印效果，如图7.6所示。

图 7.6　默认日志 Level 的效果

修改配置，设置值为%-5p，配置代码如下：

```
logging:
  pattern:
    level: "%-5p"
```

运行程序，观察会发现日志级别修改成了左对齐，在右侧补充空格，日志打印如图7.7所示。

图 7.7　%-5p 的日志 Level 效果

修改配置文件，将值设置为%p，配置代码如下：

```
logging:
  pattern:
    level: "%p"
```

运行程序，观察日志，发现已经没有对齐，日志打印如图7.8所示。

图 7.8　%p 的日志 Level 效果

上面的配置中，5的含义为不少于5个字符，如果少于5个字符，则补充空格。如果将值修改成%3p，那么效果看起来和%p一样，因为没有少于4个字符的日志级别代码。如果将值修改为%20p，那么将会出现许多空白字符，效果如图7.9所示。

图 7.9　%20p 的日志 Level 效果

再来看另一个配置项logging.pattern.dateformat，是关于日期和时间格式的，默认值是yyyy-MM-dd HH:mm:ss.SSS，从上面的截图中也可以看出显示效果，默认精确到了毫秒，通常也无须修改。

7.2.7 输出到控制台

Spring Boot默认配置日志输出包含输出到控制台，并且输出的日志级别在INFO级别以上（即INFO、WARN和ERROR）。

可以通过日志级别的配置项"logging.level.*"来配置输出到控制台的日志级别。也可以通过启动参数来修改日志级别，比如使用"--debug"或"--trace"，代码如下：

```
java -jar demo.jar --debug
```

或者

```
java -jar demo.jar --trace
```

也可以在Spring Boot的配置文件中指定，配置代码如下：

```
debug=true
```

或者

```
trace=true
```

关于控制台输出的日志格式和内容，通过配置项logging.pattern.console来设置。

通过在配置文件中修改spring.output.ansi.enabled来启用日志的颜色，配置代码如下：

```
spring:
  output:
    ansi:
      enabled: "ALWAYS"
```

运行测试代码，可以看到日志输出已经有了颜色（请读者自行验证），如图7.10所示。

```
2021-08-10 17:09:10.605 TRACE 2574 --- [          main] com.example.demo.DemoApplicationTests    : 测试日志输出 trace
2021-08-10 17:09:10.606 DEBUG 2574 --- [          main] com.example.demo.DemoApplicationTests    : 测试日志输出 debug
2021-08-10 17:09:10.606  INFO 2574 --- [          main] com.example.demo.DemoApplicationTests    : 测试日志输出 info
2021-08-10 17:09:10.606  WARN 2574 --- [          main] com.example.demo.DemoApplicationTests    : 测试日志输出 warn
2021-08-10 17:09:10.608 ERROR 2574 --- [          main] com.example.demo.DemoApplicationTests    : 测试日志输出 error
```

图7.10 控制台日志颜色输出

默认配置下，不同的日志级别标识的颜色不同，具体列举在表7.3中。

表7.3 日志级别标识的默认颜色

日志级别	默认颜色	日志级别	默认颜色
FATAL	红色	INFO	绿色
ERROR	红色	DEBUG	绿色
WARN	黄色	TRACE	绿色

还可以通过标志符"%clr"和在花括号中指定颜色关键字,来为日志的一部分或全部指定颜色,比如将时间指定为蓝色,配置代码如下:

```
logging:
  pattern:
    console: "%clr(%d{${LOG_DATEFORMAT_PATTERN: yyyy-MM-ddHH:mm:ss.SSS}}){blue} %clr(${LOG_LEVEL_PATTERN:%5p}) %clr(${PID:-}){magenta} %clr(---){faint} %clr([%15.15t]){faint}%clr(%-40.40logger{39}){cyan}%clr(:){faint}%m%n${LOG_EXCEPTION_CONVERSION_WORD:%wEx}"
```

运行测试代码,日期和时间打印成了蓝色(请读者自行验证),如图7.11所示。

```
2021-08-1017:33:54.617  INFO 2716 ---[           main]com.example.demo.DemoApplicationTests    :No active profile se
2021-08-1017:33:55.488  INFO 2716 ---[           main]com.example.demo.DemoApplicationTests    :Started DemoApplica
2021-08-1017:33:55.783 TRACE 2716 ---[           main]com.example.demo.DemoApplicationTests    :测试日志输出 trace
2021-08-1017:33:55.784 DEBUG 2716 ---[           main]com.example.demo.DemoApplicationTests    :测试日志输出 debug
2021-08-1017:33:55.785  INFO 2716 ---[           main]com.example.demo.DemoApplicationTests    :测试日志输出 info
2021-08-1017:33:55.785  WARN 2716 ---[           main]com.example.demo.DemoApplicationTests    :测试日志输出 warn
2021-08-1017:33:55.785 ERROR 2716 ---[           main]com.example.demo.DemoApplicationTests    :测试日志输出 error
```

图 7.11　日志日期时间部分设置为蓝色

注意

使用配置项logging.pattern.dateformat无法为日期时间设置颜色,不恰当的使用还会造成默认日志的混乱。

支持的颜色关键字列举在表7.4中。

表 7.4　日志颜色关键字

颜色关键字	颜　　色	颜色关键字	颜　　色
blue	蓝色	magenta	品红色
cyan	蓝绿色	red	红色
faint	默认颜色	yellow	黄色
green	绿色		

7.2.8　日志组

在配置Logger相关信息时需要指定Logger的name,如果要配置多个Logger,即便配置内容相同,也需要重复配置多次。

为了减少配置,减少不必要的重复配置,Spring Boot允许通过Spring Environment将多个Logger归并到日志组下,然后使用日志组的名称代替Logger的name,实现一次配置多个Logger的效果。

Spring Boot预定义了两个日志组:web和sql。

- 日志组 web 包括的 Logger name 有 org.springframework.core.codec、org.springframework.http、org.springframework.web、org.springframework.boot.actuate.endpoint.web 和 org.springframework.boot.web.servlet.ServletContextInitializerBeans。
- 日志组 sql 包括的 Logger name 有 org.springframework.jdbc.core、org.hibernate.SQL 和 org.jooq.tools.LoggerListener。

也可以通过配置项logging.group.*来定义日志组。配置项中*的位置为日志组的名称，值为Logger name，多个时使用英文逗号分隔。

【示例7.5】 自定义日志组

自定义日志组名称为test，包括自己代码的日志和Spring框架的日志，具体配置代码如下：

```
logging:
  group:
    test: "com.demo,org.springframework"
```

7.3 输出到日志文件的配置

Spring Boot默认配置不会输出到日志文件，可以通过配置输出到日志文件，以及配置日志滚动策略等。本节将介绍输出到日志文件的相关配置。

7.3.1 配置输出到日志文件

2.5版本的Spring Boot可以通过两个配置项输出到日志文件，这两个配置项容易被误解，这两个配置项分别是logging.file.name和logging.file.path。

从这两个配置项的命名上，可能容易被理解为配置项logging.file.path指定日志文件的存放路径，另一个配置项logging.file.name指定日志文件名。其实这两个参数并不是这么使用的。

这两个配置项没有默认值，也不会输出日志到日志文件。这两个配置项只要指定一个，即可开启向日志文件的输出功能。它们的功能不相同，并且同时指定两个时只有一个配置生效。

配置项logging.file.name用来指定日志文件名，可以包括路径，路径可以是绝对路径，也可以是相对路径。不指定路径时，如果在IDE中运行，日志文件会被写在项目根目录下，如果通过命令"java -jar"来执行，那么日志文件会在执行命令的目录下创建。

【示例7.6】 指定日志文件名

指定带绝对路径的日志文件名，配置代码如下：

```
logging:
  file:
    name: "/var/log/test/abc.log"
```

执行测试程序，然后查看系统文件，可以看到在"/var/log/test"目录下生成了日志文件abc.log，如图7.12所示。

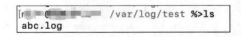

图7.12 生成的日志文件 abc.log

查看文件abc.log的内容，可以看出正是在测试方法中打印的日志输出，如图7.13所示。

另一个配置项logging.file.path只是定义日志的路径，其值被认定为日志路径，如果指定值为/var/log/test/abc.log，那么abc.log会被当作目录名，并创建目录abc.log，日志文件使用默认的日志名称spring.log。

图 7.13　日志文件 abc.log 的内容

【示例7.7】 指定配置项logging.file.path

不妨修改上一个示例中的配置项，将logging.file.name修改为logging.file.path，那么此时的配置如下：

```
logging:
  file:
    path: "/var/log/test/abc.log"
```

再次执行代码，并查看文件系统，可以看到在/var/log/test目录下出现了一个新目录abc.log，如图7.14所示。

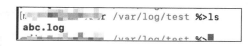

图 7.14　目录为 abc.log

注意

在/var/log/test目录下有文件名为abc.log的文件时不会创建同名的目录。

进入目录abc.log中，并查看文件列表，可以看见目录中有日志文件spring.log，最后查看日志文件内容，如图7.15所示。

图 7.15　abc.log 目录中的日志文件和文件内容

7.3.2　日志滚动配置

程序稳定运行之后，可能几个星期甚至几个月才会重启一次，这么长的运行周期，如果日志始终输入一个日志文件中，日志文件会非常大，既浪费存储空间又不方便根据日志文件查找问题，而解决这些问题的方式之一就是日志滚动策略。

如果使用Logback作为日志实现框架，那么可以在Spring Boot的配置文件中对日志滚动策略进行微调，不过如果是比较大的调整，还是需要在具体实现框架的配置文件中进行设置。

根据7.2.4节的配置项汇总，可以看到，Spring Boot提供了5个用来配置日志滚动策略的配置项。下面逐一来进行说明。

- logging.logback.rollingpolicy.file-name-pattern：用来配置日志滚动时的文件命名规则。默认值为${LOG_FILE}.%d{yyyy-MM-dd}.%i.gz，其含义是指产生新的日志文件时，将旧的日志以日期命名，如果同一天内有多个文件，再以数字为序号标注。
- logging.logback.rollingpolicy.max-file-size：用来指定单个日志文件的最大大小，当单个文件超过这一大小时，就创建新的文件，默认值为10MB。
- logging.logback.rollingpolicy.max-history：默认值为7，表示打包后的日志最多被保留7天。
- logging.logback.rollingpolicy.clean-history-on-start：指定是否在项目启动时清理历史日志文件，默认值为false。
- logging.logback.rollingpolicy.total-size-cap：保留的所有日志的总大小。当超出设置的大小后，就删除历史日志文件。

7.4 配置文件扩展

对于Logback，如果使用独立的配置文件，有带"-spring"的文件名可选，这些带"-spring"的配置文件会交由Spring解析，Spring对原生的配置文件做了一些拓展。本节将简要介绍这些拓展。

7.4.1 定义 Profile 的个性配置

通常测试、开发和生产等不同的Profile会定义不同的配置，日志输出等级、格式、内容等可能也会有根据Profile来配置的需求，此需求就可以通过Spring Boot提供的<springProfile>标签来实现。

在包含"-spring"命名的配置文件中将定义不同Profile的配置，并将不同Profile的配置使用<springProfile>标签包裹起来，并且标签<springProfile>的name属性可以通过SpEL表达式，指定在某些Profile下生效或者不生效。

具体使用可以参考如下配置代码和注解：

```
<springProfile name="dev">
    <!-- 这里的配置只会在 Profile 为 dev 时生效 -->
</springProfile>

<springProfile name="dev| test">
    <!-- 这里的配置会在 Profile 为 dev 或 test 时生效 -->
</springProfile>

<springProfile name="!prod">
    <!-- 这里的配置在 Profile 不是 prod 时生效 -->
</springProfile>
```

7.4.2 引入 Spring Environment Property

通过Spring Boot对配置文件的解析,还支持在日志框架的配置文件(-spring)中使用Spring Environment Property。通过标签<springProperty>实现对Spring Boot配置文件中变量(属性)的读取。

标签<springProperty>是对Logback标签<property>的增强,<springProperty>支持的属性有4个,分别是scope、name、source和defaultValue。

scope也是Logback的<Property>支持的属性,有3个值可选,分别是local、context和system。local表示仅在当前配置文件中可以使用,context可以在上下文过程中使用,system将会设置到JVM的系统属性中。

name和source搭配使用,比如Spring Boot配置文件中的属性logging.file.name,这里source就是logging.file,name就是name。

最后的属性defaultValue为变量定义的默认值。

【示例7.8】 在logback-spring.xml中使用<springProperty>

Spring Boot的配置文件中定义了属性test.nodes,具体代码如下:

```
test:
  nodes:
    nodeId: "node1"
```

logback-spring.xml中使用标签<springProperty>取值,设置scope为context,设置source和name并使用,具体代码如下:

```
<configuration>

  <springProperty scope="context" name="nodeId" source="test.nodes" />

  <appender name="FILE" class="ch.qos.logback.core.FileAppender">
    <file>/opt/${nodeId}/myApp.log</file>
    <encoder>
      <pattern>%msg%n</pattern>
    </encoder>
  </appender>

  <root level="debug">
    <appender-ref ref="FILE" />
  </root>
</configuration>
```

第 8 章

Spring Boot的安全与监控

当项目发布到生产环境后，必须考虑项目的访问安全，也就是需要进行访问限制，Spring Boot对安全框架Spring Security进行了默认配置。此外，生产环境，尤其是分布式或服务集群，通常需要监控程序的运行状态，这就是所谓的监控。

本章主要涉及的知识点有：

- 如何使用Spring Security。
- Actuator是什么。
- Actuator提供了哪些功能。
- 如何自定义Endpoint。

8.1 安全控制（使用Spring Security）

系统的用户认证和授权通常是项目尤其是Web项目必不可少的功能，在第3章介绍使用Spring Boot开发Web项目时，涉及使用安全框架Shiro实现用户认证的功能。本节将介绍在Spring Boot项目中对安全框架Spring Security的开启和使用。

8.1.1 Spring Security 的开启和配置

Spring Boot对Spring Security做了默认配置，在类路径中存在Spring Security时会自动启用，所以在Spring Boot项目中启用Spring Security只需要引入相关的依赖。Spring Boot为Spring Security提供Starter，进而开启Spring Security只需要在pom.xml中添加对spring-boot-starter-security的依赖，具体配置代码如下：

```
<dependency>
    <groupId>org.springframework.boot</groupId>
    <artifactId>spring-boot-starter-security</artifactId>
</dependency>
```

刷新Maven的依赖，然后启动项目，可以看到控制台输出的日志中出现了Spring Security相关的内容，如图8.1所示。

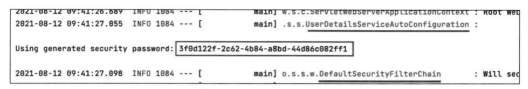

图 8.1　输出 user 用户的随机密码

日志中输出了一个随机密码，这是因为在自动配置类UserDetailsServiceAutoConfiguration时生成了一个默认用户，用户名为user，密码为使用UUID生成的一个字符串，正如图8.1中所示的3f0d122f-2c62-4b84-a8bd-44d86c082ff1。

【示例8.1】　使用默认用户登录

在浏览器中访问地址http://127.0.0.1:8080/，自动重定向到http://127.0.0.1:8080/login，并出现用户名和密码的登录页面，具体如图8.2所示。

图 8.2　默认的登录页面

输出默认用户名user，并使用默认密码3f0d122f-2c62-4b84-a8bd-44d86c082ff1，登录成功后会跳转至http://127.0.0.1:8080/，不过由于项目并没有配置页面，因此会显示Whitelabel Error Page，如图8.3所示，这是正常的。

接下来查看一下请求如何携带登录信息。

在页面上右击，在弹出的菜单中选择"检查"，如图8.4所示。

图 8.3　登录后的默认 404 页面

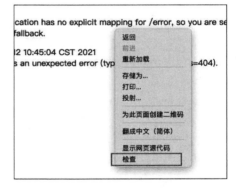

图 8.4　Chrome 右键菜单

在打开的开发者工具窗口（DevTools）中切换到Network标签页，如图8.5所示。

然后在浏览器上刷新页面，开发者工具窗口中会出现网络请求的记录，如图8.6所示。请求响应404是因为当前项目中没有配置请求的路径，并不会影响登录状态。

单击选中请求，并在出现的Headers标签页中展开Request Headers部分，如图8.7所示，可以看到请求头中包含Cookie这一项。这是Spring Security默认使用Cookie技术记录了客户端的登录状态。

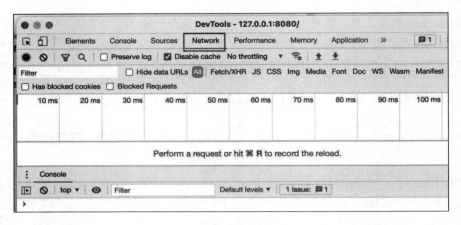

图 8.5　Chrome 开发者工具

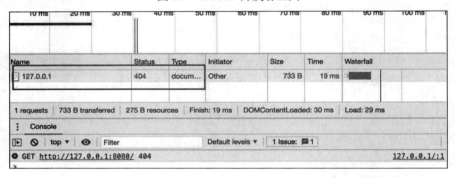

图 8.6　请求记录

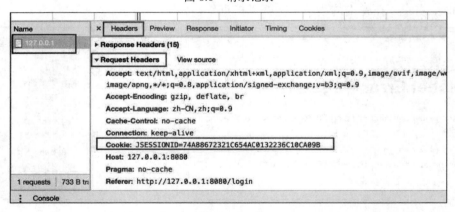

图 8.7　请求头的 Cookie 信息

前面提到一个Spring Boot默认的配置就是默认用户user和随机密码，这个用户信息是在UserDetailsServiceAutoConfiguration类中配置的。此外，Spring Boot的自动配置还会创建一个name为springSecurityFilterChain的Filter Bean，默认会拦截所有请求，并重定向到/login页面，也就是在上面的示例中看到的页面。

除了登录页面外，Spring Security还默认配置了一个退出页面。在浏览器中访问地址http://127.0.0.1:8080/logout，会进入退出页面，如图8.8所示。

单击Log Out按钮，退出登录，并且跳转回登录页面（http://127.0.0.1:8080/login）。

此外，还有一些默认配置，比如使用BCrypt作为密码加密算法，阻止跨域请求和会话固定保护等。

8.1.2 使用 Spring Security

默认配置通常不能满足我们的需求，因此需要我们做一些自定义的设置。

图 8.8　默认退出页面

1. 自定义配置类

使用代码进行配置，首先需要创建一个配置类，使用注解@Configuration标注为配置类，并继承WebSecurityConfigurerAdapter。例如，可以在config包下创建security包，并将后面所有需要的安全相关的配置类都放置于此包下。比如，Spring Security将配置类命名为SecurityConfig，代码如下：

```java
package com.example.demo.config.security;

import org.springframework.context.annotation.Configuration;
import org.springframework.security.config.annotation.web.configuration.WebSecurityConfigurerAdapter;

@Configuration
public class SecurityConfig extends WebSecurityConfigurerAdapter {}
```

2. 自定义用户认证策略

默认实现中，用户认证是通过UserDetailsService这个接口来实现的，前面提到的默认用户user就是在这个接口的默认实现类中配置的。不过默认是通过session实现的，而目前更多使用token实现，这里的实现可以不使用UserDetailsService，而采用登录接口返回token并配合验证token过滤器来实现。

首先要自定义User实体类，保存用户的用户名和密码等基本信息，可以参考如下代码：

```java
package com.example.demo.entity;

import javax.persistence.Entity;
import javax.persistence.Id;
import javax.persistence.Table;

@Entity
@Table
public class User{
    private String id;
    private String username;
    private String password;
    // 省略了构造器、getter、setter、equals、hashcode 等
}
```

还需要对 User 数据访问的接口，这里以用户数据存储在数据库为例，并且采用 spring-data-jpa 的方式来访问数据。访问用户数据通过 UserDao 来实现，UserDao 接口的代码如下：

```
package com.example.demo.dao;

import com.example.demo.entity.User;
import org.springframework.data.jpa.repository.JpaRepository;
import org.springframework.data.jpa.repository.JpaSpecificationExecutor;

public interface UserDao extends JpaRepository<User, String>,
JpaSpecificationExecutor<User> {
}
```

接下来需要自定义登录接口来替换掉默认的登录实现，并配置自定义的登录接口免登录访问，以及需要登录接口返回 token、token 的创建和过滤器校验等。接下来分步实现，并在最后将这些串起来。

3. 自定义的 token 工具类

自定义的 token 工具类应该具有根据 User 实例创建 token、从 token 中解析出用户 ID、校验 token 是否过期等功能。这里创建 token 和解析 token 采用 jjwt 工具来实现，所以需要先引入 jjwt 的依赖，配置代码如下：

```xml
<dependency>
    <groupId>io.jsonwebtoken</groupId>
    <artifactId>jjwt</artifactId>
    <version>0.9.0</version>
</dependency>
```

创建 token 工具类，TokenUtil 类代码如下所示：

```java
package com.example.demo.config.security;

import com.example.demo.entity.User;
import io.jsonwebtoken.Claims;
import io.jsonwebtoken.Jwts;
import io.jsonwebtoken.SignatureAlgorithm;
import org.springframework.stereotype.Component;

import java.util.Date;
import java.util.HashMap;
import java.util.Map;

@Component
public class TokenUtil {
    private static final String CLAIM_KEY_USER_ID = "userId";
    private static final String CLAIM_KEY_CREATED = "created";
    private static final String secret = "xez3331sdf2fd23ske";
    private static final long expiration = 7*24*60*60*1000L; // token 指定过期时间为 7 天

    /**
     * 根据用户信息生成 Token
     *
```

```java
 * @return
 */
public String generateToken(User user) {
    Map<String, Object> claims = new HashMap<>();
    claims.put(CLAIM_KEY_USER_ID, user.getId());
    claims.put(CLAIM_KEY_CREATED, new Date());
    return generateToken(claims);
}

private String generateToken(Map<String, Object> claims) {
    return Jwts.builder().setClaims(claims).setExpiration(genExpirationDate())
            .signWith(SignatureAlgorithm.HS256, secret).compact();
}

private Date genExpirationDate() {
    return new Date(System.currentTimeMillis() + expiration);
}

/**
 * 从 Token 中获取用户名
 *
 * @param token
 * @return
 */
public String getUserIdFromToken(String token) {
    String userId;
    try {
        Claims claims = getClaimsFromToken(token);
        userId = claims.getSubject();
    } catch (Exception e) {
        return null;
    }
    return userId;
}

/**
 * 从 token 中获取荷载
 *
 * @param token
 * @return
 */
private Claims getClaimsFromToken(String token) {
    Claims claims = null;
    try {
        claims = Jwts.parser().setSigningKey(secret)
                .parseClaimsJws(token)
                .getBody();
    } catch (Exception e) {
        e.printStackTrace();
    }
    return claims;
```

```java
    }
    /**
     * 判断token是否有效 && 是否过期
     *
     * @return
     */
    public boolean validateToken(String token, User user) {
        // 判断是否过期,判断是否有效
        String userId = getUserIdFromToken(token);
        return userId.equals(user.getId())
                && !isTokenExpired(token);
    }

    public boolean isTokenExpired(String token) {
        Date expireDate = getExpireDateFromToken(token);
        return expireDate.before(new Date());
    }

    public Date getExpireDateFromToken(String token) {
        Claims claims = getClaimsFromToken(token);
        return claims.getExpiration();
    }

    /**
     * 判断token是否可以被刷新
     *
     * @param token
     * @return
     */
    public boolean canRefresh(String token) {
        return !isTokenExpired(token);
    }

    public String refreshToken(String token) {
        Claims claims = getClaimsFromToken(token);
        claims.put(CLAIM_KEY_CREATED, new Date());
        return generateToken(claims);
    }
}
```

4. 自定义登录接口

登录接口需要做的是接收用户名和密码,并对用户名和密码做校验。这里接收到的密码通常是明文密码,然而存储在数据库中的密码通常是密文,所以这里还需要对密码进行加密和校验的工作。以使用BCrypt为例,首先在容器中添加BCrypt的加密器,可以在SecurityConfig中定义,参考如下代码:

```java
@Bean
PasswordEncoder passwordEncoder(){
    return new BCryptPasswordEncoder();
}
```

接下来定义登录接口，使用POST请求，为了安全起见，将用户名和密码放在请求体中，以JSON格式传递，接口参数使用实体类接收。大致思路是根据接收到的username，使用UserDao从数据库中查询用户信息，然后将查询到的用户信息的password和接收到的password做校验，如果所有校验通过，则认定成功登录，并返回token。创建LoginController，并实现校验逻辑，具体代码如下：

```java
package com.example.demo.controller;

import com.example.demo.config.security.TokenUtil;
import com.example.demo.dao.UserDao;
import com.example.demo.entity.User;
import org.springframework.beans.factory.annotation.Autowired;
import org.springframework.security.crypto.password.PasswordEncoder;
import org.springframework.web.bind.annotation.PostMapping;
import org.springframework.web.bind.annotation.RequestBody;
import org.springframework.web.bind.annotation.RequestMapping;
import org.springframework.web.bind.annotation.RestController;

import java.util.HashMap;
import java.util.Map;

@RestController
@RequestMapping("login")
public class LoginController {
    @Autowired
    UserDao userDao;
    @Autowired
    TokenUtil tokenUtil;
    @Autowired
    PasswordEncoder passwordEncoder;
    @PostMapping("bypassword")
    public Map loginByPassword(@RequestBody User userLogin){
        User user = userDao.findOneByUsername(userLogin.getUsername());
        return updateSecurityUserInfo(user);
    }
    public Map updateSecurityUserInfo(User user){
        Map<String,Object> retMap = new HashMap<>();
        String token = tokenUtil.generateToken(user);
        retMap.put("token", token);
        return retMap;
    }
}
```

5. 创建 token 过滤器

自定义一个Filter，需要在这个过滤器中校验token，并将token设置到请求相关的安全上下文中，以便在业务逻辑中方便取出登录用户的信息。在config.security包下创建Filter的实现类AuthencationTokenFilter，为了方便实现，直接实现OncePerRequestFilter类，这样间接继承了Filter接口，并在doFilterInternal方法中实现具体的逻辑。具体实现代码如下：

```java
package com.example.demo.config.security;

import com.example.demo.dao.UserDao;
import com.example.demo.entity.User;
import org.springframework.beans.factory.annotation.Autowired;
import org.springframework.orm.jpa.JpaObjectRetrievalFailureException;
import org.springframework.security.authentication.UsernamePasswordAuthenticationToken;
import org.springframework.security.core.context.SecurityContextHolder;
import org.springframework.security.web.authentication.WebAuthenticationDetailsSource;
import org.springframework.web.filter.OncePerRequestFilter;
import javax.servlet.FilterChain;
import javax.servlet.ServletException;
import javax.servlet.http.HttpServletRequest;
import javax.servlet.http.HttpServletResponse;
import java.io.IOException;

public class AuthencationTokenFilter extends OncePerRequestFilter {
    @Autowired
    TokenUtil tokenUtil;
    @Autowired
    UserDao userDao;

    @Override
    protected void doFilterInternal(HttpServletRequest request, HttpServletResponse response, FilterChain filterChain) throws ServletException, IOException {
        String authToken = request.getHeader("Authorization");
        String userId = tokenUtil.getUserIdFromToken(authToken);
        // 可以从 Token 中拿到用户信息,但是上下文中尚未设置用户信息
        if (null != userId && null == SecurityContextHolder.getContext().getAuthentication()) {
            User user = null;
            try {
                user = userDao.getById(userId);
            } catch (JpaObjectRetrievalFailureException e) {
                System.out.println("token 中的 userId 无效" + userId);
                return;
            }
            if (user != null && tokenUtil.validateToken(authToken, user)) {
                UsernamePasswordAuthenticationToken usernamePasswordAuthenticationToken = new UsernamePasswordAuthenticationToken(user, null, null);
                usernamePasswordAuthenticationToken.setDetails(new WebAuthenticationDetailsSource().buildDetails(request));
                SecurityContextHolder.getContext().setAuthentication(usernamePasswordAuthenticationToken);
            } else {
                System.out.println("token 中的 userId 无效" + userId);
                return;
```

```
            }
        }
        filterChain.doFilter(request, response);
    }
}
```

校验失败时,为了给接口调用方提示信息,可以在return之前通过输出流的形式输出提示信息。

创建的AuthencationTokenFilter还需要为其创建实例,并将实例添加到Spring容器中,这一过程同样可以通过配置类和@Bean注解来实现。在SecurityConfig类中添加配置代码,具体代码如下:

```
@Bean
public AuthencationTokenFilter jwtAuthencationTokenFilter(){
    return new AuthencationTokenFilter();
}
```

然后还需要将其配置给Spring Security组件,使用WebSecurityConfigurerAdapter类提供的方法configure(HttpSecurity http)。SecurityConfig类已经实现了WebSecurityConfigurerAdapter,所以只需要在类SecurityConfig中添加方法configure(HttpSecurity http),并在方法中配置filter,具体代码如下:

```
@Override
protected void configure(HttpSecurity http) throws Exception {
    // 添加登录授权过滤器
    http.addFilterBefore(authencationTokenFilter(), UsernamePasswordAuthenticationFilter.class);
}
```

6. 配置免登录放行规则

在上面完成了登录接口、token的生成和使用拦截器对token校验之后,接下来需要配置来放行免登录的接口,比如登录接口不可以做拦截。

要配置的访问拦截规则相对简单,只需要设置登录接口免登录访问,其他接口都是需要登录才能访问,注意配置的顺序也是从具体到通用。这里配置仍然用到WebSecurityConfigurerAdapter类的方法。WebSecurityConfigurerAdapter类中有多个configure重载方法,在上面配置filter用到的configure(HttpSecurity http)或者重载方法configure(WebSecurity webSecurity)都可以完成拦截规则配置。在这里仍然使用configure(HttpSecurity http)来举例,需要在该方6CD5中添加以下代码:

```
http.authorizeRequests().antMatchers(HttpMethod.POST, "/login/**").
permitAll().and().authorizeRequests().anyRequest().authenticated();   // 其他需
要登录
```

通过这个方法提供的形参HttpSecurity http还可以进行其他配置,比如禁用跨域限制、弃用session策略和不使用缓存等,代码如下:

```
http.csrf().disable() // 使用jwt,不用csrf,解决跨域问题
        .sessionManagement().sessionCreationPolicy(SessionCreationPolicy.STATELESS) // 不使用session
        .and().headers().cacheControl();// 用不到缓存
```

最后，如果项目开启了Actuator，那么安全控制还需要对Actuator控制，具体使用将在下一节介绍。

8.2 使用Actuator监控应用

Spring Boot Actuator是Spring Boot提供的监控和管理应用的一系列功能合集。Spring Boot Actuator按功能划分成Endpoint，每个功能对应一个Endpoint。本节将介绍Spring Boot Actuator默认提供的Endpoints，以及如何自定义Endpoint。

8.2.1 开启 Actuator

spring-boot-actuator是实现Spring Boot Actuator的项目，Spring Boot为spring-boot-actuator提供了Starter。所以要在Spring Boot项目中开启Actuator，只需要引入spring-boot-starter-actuator依赖，配置代码如下：

```
<dependencies>
    <dependency>
        <groupId>org.springframework.boot</groupId>
        <artifactId>spring-boot-starter-actuator</artifactId>
    </dependency>
</dependencies>
```

为了校验引入Actuator的状态，先将Spring Security配置成允许所有URL免登录访问，在SecurityConfig类中修改配置代码，具体代码如下：

```
http.authorizeRequests().antMatchers("/**").permitAll()
```

启动项目，在浏览器中访问http://127.0.0.1:8080/actuator，这是Actuator的发现页，如图8.9所示，说明访问成功。

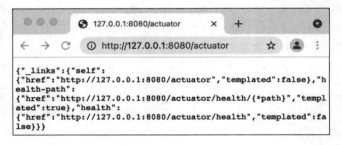

图 8.9　Actuator 的发现页

进一步访问Health Endpoint，在浏览器中访问 http://127.0.0.1:8080/actuator/health，效果如图8.10所示。

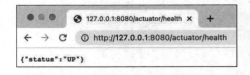

图 8.10　Health Endpoint 默认响应信息

8.2.2 默认配置

Actuator按功能划分成一个个独立的Endpoint，Endpoint类似于接口，也是通过Endpoint实现对项目的监控和管理。

spring-boot-actuator内置了一些Endpoint并支持开发者自定义Endpoint。

每个Endpoint都可以独立地被开启（Enabled）或者禁用（Disabled）。对于开启的Endpoint，又可以设置其是否允许被访问（Exposed），只有开启并允许被访问时，才称之为可用（Available）。

每个Endpoint都可以设置其被访问的实现技术，有HTTP和JMX可选。对于大多数应用程序，尤其是Web程序，都会选择HTTP的访问方式。在上一节，测试Actuator状态时，在浏览器中使用地址http://127.0.0.1:8080/actuator和http://127.0.0.1:8080/actuator/health进行访问就是通过HTTP的方式。

关于spring-boot-actuator内置的Endpoint，这些Endpoint默认被开启以及所采用的访问技术汇总信息参见表8.1。

表 8.1 内置的 Endpoint 及默认配置汇总

Endpoint ID	Enabled	Exposed		说 明
		JMX	HTTP	
auditevents	是	是	否	显示当前应用的审计事件的信息，在容器中有AuditEventRepository类型的实例时才会开启
beans	是	是	否	显示应用程序的Spring容器中的所有实例
caches	是	是	否	显示可用的缓存
conditions	是	是	否	显示所有自动配置评估结果和匹配或者不匹配的原因
configprops	是	是	否	展示所有@ConfigurationProperties的列表
env	是	是	否	展示Spring所有的Environment Property
flyway	是	是	否	显示已应用的所有Flyway数据库迁移。在具有一个或多个Flyway实例时可用
health	是	是	是	展示应用的健康信息
httptrace	是	是	否	展示HTTP请求响应信息，默认展示最新的100条。在有HttpTraceRepository实例时可用
info	是	是	否	展示应用程序的多种信息
integrationgraph	是	是	否	显示Spring的依赖图。在依赖了项目spring-integration-core时可用
loggers	是	是	否	展示和修改Logger的配置
liquibase	是	是	否	显示已应用的任何Liquibase数据库迁移。在具有Liquibase实例时可用
metrics	是	是	否	展示应用程序的metrics信息
mappings	是	是	否	展示所有@RequestMapping的地址
quartz	是	是	否	展示Quartz Scheduler任务的信息
scheduledtasks	是	是	否	展示程序中的定时任务

（续表）

Endpoint ID	Enabled	Exposed JMX	Exposed HTTP	说明
sessions	是	是	否	允许从Spring Session支持的会话存储中检索和删除用户会话。在使用Spring Session的Servlet Web应用程序中可用
shutdown	否	是	否	让应用程序自行停止。默认关闭了此Endpoint
startup	是	是	否	展示出由ApplicationStartup收集的启动数据。在BufferingApplicationStartup配置SpringApplication中可用
threaddump	是	是	否	对指定的线程dump
heapdump	是	不可用	否	返回hprof的heap dump文件，必须是HotSpot JVM
jolokia	是	不可用	否	通过HTTP暴露JMX。只有当Jolokia在类路径下生效，并且不支持WebFlux
logfile	是	不可用	否	返回日志文件的内容
prometheus	是	不可用	否	以Prometheus服务的形式曝光metrics。在依赖micrometer-registry-prometheus时可用

列表最后的4个Endpoint，即heapdump、jolokia、logfile和prometheus，只有在应用是Web应用，即使用了Spring MVC、Spring WebFlux或Jersey技术时才可以使用。

默认条件下，除了shutdown外，其他Endpoint都是启用的。可以在Spring的配置文件中通过management.endpoints.enabled-by-default设置所有Endpoint的默认状态，比如可以将该值设置为false，将所有Endpoint默认设置为禁用，具体配置代码如下：

```
management:
  endpoints:
    enabled-by-default: false
```

设置为禁用的Endpoint将会从应用程序上下文中完全移除。此外，也可以通过management.endpoint.<id>.enabled属性，为ID是<id>的Endpoint单独设置是否启用。比如对于shutdown endpoint，可以通过如下配置来启用：

```
management:
  endpoint:
    shutdown:
      enabled: true
```

不过现在并不能通过HTTP访问shutdown，因为HTTP方式默认只开放了health。

Spring Boot提供了4个属性配置内置Endpoint的开放策略，management.endpoints.jmx.exposure.exclude和management.endpoints.jmx.exposure.include用来配置Endpoint是否通过JMX方式开放。默认设置下，jmx开放所有Endpoint。management.endpoints.web.exposure.exclude和management.endpoints.web.exposure.include用来配置Endpoint是否通过HTTP方式开放。

默认配置下，management.endpoints.web.exposure.include的值为health。如果需要开放health和shutdown，那么应该指定此值为"health,shutdown"，允许shutdown访问以及开启shutdown的配置代码如下：

```yaml
management:
  endpoint:
    shutdown:
      enabled: true
  endpoints:
    web:
      exposure:
        include: "health,shutdown"
```

修改配置后重启项目，然后在浏览器中访问http://127.0.0.1:8080/actuator，页面显示已经开启了shutdown，并且给出了shutdown endpoint的访问地址，如图8.11所示。

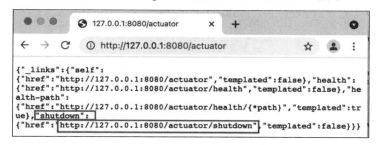

图 8.11　开启 shutdown 的 Actuator 发现页

注意，如果此时直接在浏览器地址栏访问http://127.0.0.1:8080/actuator/shutdown，那么浏览器会以GET方法发送请求，结果如图8.12所示。

图 8.12　无权限访问

正确的方式是发送POST请求，而POST请求可以通过POST MAN工具发送。这里通过命令行使用curl命令，命令代码如下：

```
curl -X POST "http://127.0.0.1:8080/actuator/shutdown"
```

发送命令并接收到响应，如图8.13所示。

```
~ %>curl -X POST  "http://127.0.0.1:8080/actuator/shutdown"
{"message":"Shutting down, bye..."}%
```

图 8.13　CURL 发送 HTTP 请求

可以看到应用程序正在关闭，然后查看IDE控制台，上面信息显示程序关闭的过程以及最终完全关闭，如图8.14所示。

```
2021-08-14 13:18:24.159  INFO 1806 --- [       Thread-1] j
.LocalContainerEntityManagerFactoryBean  : Closing JPA EntityManagerFactory
 for persistence unit 'default'
2021-08-14 13:18:24.163  INFO 1806 --- [       Thread-1] c.z.h
.HikariDataSource                        : HikariPool-1 - Shutdown initiated...
2021-08-14 13:18:24.180  INFO 1806 --- [       Thread-1] c.z.h
.HikariDataSource                        : HikariPool-1 - Shutdown completed.

Process finished with exit code 0
```

图 8.14 程序关闭的日志输出

【示例8.2】 禁用所有Endpoint的JMX实现

可以通过配置项management.endpoints.jmx.exposure.exclude和通配符"*"来完成设置，配置代码如下：

```
management:
  endpoints:
    jmx:
      exposure:
        exclude: "*"
```

这里要注意，星号"*"是YAML文件中的特殊字符，所以值必须使用引号引起来。

以上便是内置Endpoint的开启、允许访问和访问交互的基本设置和过程。

8.2.3 Actuator 的安全控制

前面提到为了测试方便，将Spring Security的免登录URL策略设置成了"/**"，这样的效果是允许所有HTTP接口免登录访问，但是很明显在实际项目中这么设置会使应用程序和数据面临很大的安全风险。通常根据业务需要，只会是登录相关的接口，例如"/login/*"允许免登录访问。那么如何在应用程序原有的安全设置上根据Actuator的需要增加设置，就是本节将要解决的问题。

1. Endpoint 的访问路径

在前面的内容中用到了对Actuator发现页的访问路径http://127.0.0.1:8080/actuator、Health Actuator的访问路径http://127.0.0.1:8080/actuator/health和shutdown的访问路径http://127.0.0.1:8080/actuator/shutdown。

Actuator Endpoints默认和应用程序使用同一个端口8080，也可以通过属性management.server.port进行设置。

【示例8.3】 设置Endpoints的端口使用8081

在application.yml中增加设置，配置代码如下：

```
management:
  server:
    port: 8081
```

重启项目，在浏览器中访问http://127.0.0.1:8081/actuator，页面提示404。访问8081的地址，响应正常，如图8.15所示。

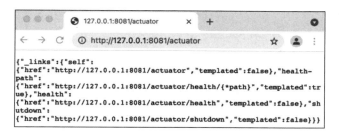

图 8.15　修改端口为 8081 后的访问效果

在IP和端口之后是Endpoint的访问路径，/actuator是默认Endpoint的根路径，也是Endpoint的发现地址，返回当前所有可访问的Endpoint和其访问地址。所有的Endpoint访问路径都是根路径拼接上/，再拼接上Endpoint的ID。例如，health的Endpoint的路径就是/actuator/health。

Endpoint的根路径可以使用配置项management.endpoints.web.base-path来修改。

【示例8.4】　将Endpoint根路径修改为/endpoint

在application.yml中增加配置，具体配置代码如下：

```
management:
  endpoints:
    web:
      base-path: "/endpoint"
```

启动项目，可以注意到控制台输出如图8.16所示，说明Endpoint的根目录设置已经生效。

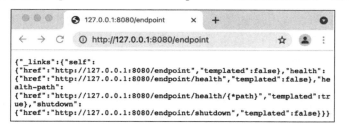

图 8.16　控制台输出 Endpoint 访问根目录

在浏览器中访问地址http://127.0.0.1:8080/endpoint，响应结果正常，响应如图8.17所示。

图 8.17　修改 Endpoint 根目录后访问

此外，还可以通过配置项management.endpoints.web.path-mapping为单个Endpoint设置访问地址，语法是management.endpoints.web.path-mapping.<endpoint id>=<path>。

【示例8.5】　将health endpoint路径指定为app-health

不妨在上一个示例的配置文件中修改，增加配置项path-mapping.health，并指定值为app-health，修改后的配置文件相关配置代码如下：

```
management:
  endpoints:
```

```yaml
    web:
      base-path: "/endpoint"
      path-mapping:
        health: "app-health"
```

重启项目，在浏览器中访问地址http://127.0.0.1:8080/endpoint，可以看到Health Endpoint的访问地址已经发生变化，如图8.18所示。

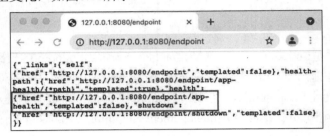

图 8.18　自定义 Health 的访问路径

访问修改后的Health Endpoint地址http://127.0.0.1:8080/endpoint/app-health，响应正常，响应如图8.19所示。

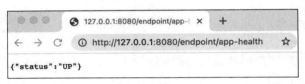

图 8.19　访问 app-health

2．配置安全策略

在了解了Endpoint的访问路径规则之后，对于Endpoint的安全策略，可以使用Spring Security提供的antMatchers方法来自定义安全策略，也可以使用Spring Boot提供的更简便的EndpointRequest.toAnyEndpoint()对象来设置安全策略。比如，可以通过如下代码只允许具有ADMIN角色的用户访问Actuator功能：

```java
@Bean
public SecurityFilterChain securityFilterChain(HttpSecurity http) throws Exception {
    http.requestMatcher(EndpointRequest.toAnyEndpoint())
        .authorizeRequests((requests) -> requests.anyRequest().hasRole("ADMIN"));
    http.httpBasic();
    return http.build();
}
```

【示例8.6】　设置Actuator的所有Endpoint允许免登录访问

先确定Spring Security中的配置，仅允许"/login"接口免登录访问，配置代码如下：

```java
http.authorizeRequests().antMatchers("/login/**").permitAll()
    .and().authorizeRequests().anyRequest().authenticated();   // 其他一定要登录
```

此时启动项目，访问http://127.0.0.1:8080/actuator，会提示403，说明配置生效，如图8.20所示。

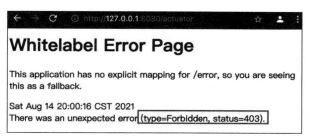

图 8.20　登录权限控制

现在需要配置所有属于Actuator的URL都允许免登录访问，大致思路是使用EndpointRequest.toAnyEndpoint()进行匹配，匹配成功的request全部都（anyRequest()）允许免登录访问（permitAll()）。

由于需要调用anyRequest()，在已有的Spring Security的配置代码中，在配置其他一定要登录时已经调用了anyRequest()，所以不能使用原先已有的configure(HttpSecurity http)方法。可以使用上面的SecurityFilterChain，或者使用Multiple HttpSecurity的配置方式，重新创建一个WebSecurityConfigurerAdapter的子类，并重写configure(HttpSecurity http)方法，在方法中定义规则，允许Actuator的URL免登录访问。新创建的安全配置类代码如下：

```
package com.example.demo.config.security;

import org.springframework.boot.actuate.autoconfigure.security.servlet.EndpointRequest;
import org.springframework.context.annotation.Configuration;
import org.springframework.core.annotation.Order;
import org.springframework.security.config.annotation.web.builders.HttpSecurity;
import org.springframework.security.config.annotation.web.configuration.WebSecurityConfigurerAdapter;

@Configuration
@Order(99)
public class ActuatorSecurityConfiguration extends WebSecurityConfigurerAdapter {
    protected void configure(HttpSecurity http) throws Exception {
        http.requestMatcher(EndpointRequest.toAnyEndpoint()).authorizeRequests((request) -> request.anyRequest().permitAll());
    }
}
```

这里有一个细节，就是Order包，不要导成log相关包下的，注意应该是org.springframework.core.annotation.Order。

配置之后重启项目，访问项目根目录http://127.0.0.1:8080/，如图8.21所示，提示403，表示没有登录，所示安全配置是生效的。

图 8.21　访问项目根目录

访问项目Actuator发现页http://127.0.0.1:8080/actuator，响应正常，如图8.22所示，说明新增加的免登录安全配置生效。

图 8.22　配置免登录后访问 Actuator 发现页

8.2.4　Health Endpoint 的使用

Health Endpoint是Spring Boot Actuator内置的Endpoint，用来检查项目的运行状态，其展示的信息和检查的项目可以设置和自定义。本节就来介绍Health Endpoint的使用。

1．配置是否展示具体信息

前面多次用到过Health Endpoint，它默认返回的信息前面几节也看到过，只是显示程序是否正在运行，即"{"status":"UP"}"。默认返回的信息不包含详细信息，可以通过配置项来修改。

配置项management.endpoint.health.show-details和配置项management.endpoint.health.show-components是等价的，都可以用来设置使health endpoint返回详细信息。

这两个配置项有3个值可选，分别是never、when-authorized和always。

- never 就是配置项的默认值。
- when-authorized 是可选值，只对具有特定角色的用户展示详细信息，而这里的角色名称可以通过配置项 management.endpoint.health.roles 来设置。
- always 表示对所有用户都展示详细信息。

【示例8.7】　设置health展示详细信息

两个配置项功能一样，这里选用management.endpoint.health.show-details，将其值指定为always，配置代码如下：

```
management:
  endpoint:
    health:
      show-details: always
```

重启项目，在浏览器中访问http://127.0.0.1:8080/actuator/health，这时返回结果中已经包含了详细信息，如图8.23所示。

图 8.23　Health Endpoint 的详情信息

具体看来，相比于默认设置下的响应结果，增加了components属性，这个属性中包括3个组件，它们是db、diskSpace和ping。

2. Health Endpoint 内置的组件

上面看到health的详细信息包含components，而components中的信息是由接口HealthContributor的实现类来提供的。HealthIndicator是HealthContributor的一个子接口，Spring Boot Actuator为HealthIndicator提供了一些内置的实现类，这些内置的实现类构成了components属性的信息。内置的HealthIndicator列举在了表8.2中。

表 8.2　内置的 HealthIndicator 实现类

key	实 现 类	描 述
cassandra	CassandraDriverHealthIndicator	检查数据库Cassandra是否可用
couchbase	CouchbaseHealthIndicator	检查Couchbase集群是否可用
db	DataSourceHealthIndicator	检查DataSource是否连接正常
diskspace	DiskSpaceHealthIndicator	检查硬盘空间的状态
elasticsearch	ElasticsearchRestHealthIndicator	检查Elasticsearch集群是否正常
hazelcast	HazelcastHealthIndicator	检查Hazelcast服务是否正常
influxdb	InfluxDbHealthIndicator	检查InfluxDB服务是否正常
jms	JmsHealthIndicator	检查JMS Broker服务是否正常
ldap	LdapHealthIndicator	检查LDAP服务是否正常
mail	MailHealthIndicator	检查邮件服务是否正常
mongo	MongoHealthIndicator	检查Mongo数据库是否正常
neo4j	Neo4jHealthIndicator	检查Neo4j数据库是否正常
ping	PingHealthIndicator	只要可以返回，就一定是UP
rabbit	RabbitHealthIndicator	检查Rabbit服务是否正常
redis	RedisHealthIndicator	检查Redis服务是否正常
solr	SolrHealthIndicator	检查Solr服务是否正常
livenessstate	LivenessStateHealthIndicator	返回Liveness应用的可用状态
readinessstate	ReadinessStateHealthIndicator	返回Readiness的可用状态

可以在列表中找到上面示例中的db、diskspace和ping。

列表中的HealthIndicator实现类除了最后的livenessstate和readinessstate外，其他都是默认启用的，只是Spring Boot会在具有相关组件时才会默认启用。

可以通过配置项management.health.defaults.enabled来统一设置所有内置的HealthIndicator启用与否。也可以通过配置项management.health.<key>.enabled来设置具体某一个HealthIndicator是否启用，这里的key对应列表的key列的值。

【示例8.8】 停用PingHealthIndicator

根据列表找到PingHealthIndicator的key是ping，所以使用配置项management.health.ping.enabled来配置，具体配置代码如下：

```
management:
    health:
        ping:
            enabled: false
```

重启程序，在浏览器中访问http://127.0.0.1:8080/actuator/health，可以看到响应结果中已经没有属性为ping的那一项，如图8.24所示。

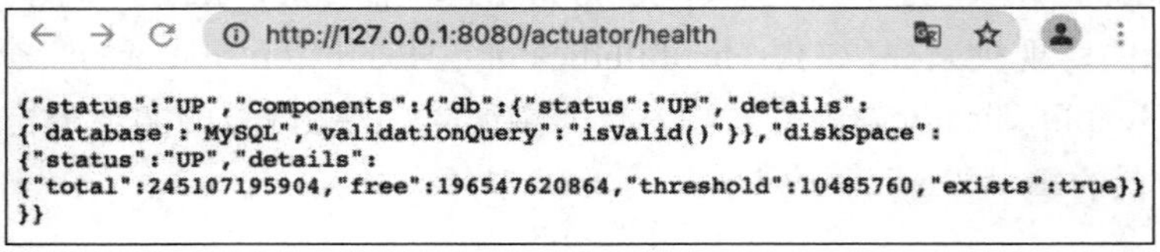

图 8.24 关闭 ping Endpoint 的访问

3. 自定义 health 组件

自定义health组件需要实现HealthIndicator接口，并重写health()方法。

health()方法返回值为Health类型。Health类型有两个字段，这两个字段分别是Status类型的status和Map<String, Object>类型的details。例如，在图8.24中可以注意到key为"db"的具体值为"{"status":"UP","details":{"database":"MySQL","validationQuery":"isValid()"}"，可以大致了解Health的组成。

自定义的类名去掉HealthIndicator的后缀，剩下的部分首字母小写，作为key。如果不存在HealthIndicator后缀，那么整个类名首字母小写后作为key。

【示例8.9】 自定义HealthIndicator

创建类CustomHealth实现HealthIndicator接口，并重写health()方法，具体代码如下：

```java
package com.example.demo.config.security;

import org.springframework.boot.actuate.health.Health;
import org.springframework.boot.actuate.health.HealthIndicator;
import org.springframework.boot.actuate.health.Status;
import org.springframework.stereotype.Component;
import java.util.HashMap;
import java.util.Map;

@Component
public class CustomHealth implements HealthIndicator {
    @Override
    public Health health() {
```

```
        Map<String, Object> map = new HashMap<>();
        map.put("custom health", "自定义的Indicator");
        return new Health.Builder(Status.UP, map).build();
    }
}
```

根据key的生成规则，customHealth将作为生成的key。代码返回的status为up，details内容为""custom health":"自定义的Indicator""。

重启项目，在浏览器中访问http://127.0.0.1:8080/actuator/health，响应的内容正符合代码中的设置，具体如图8.25所示。

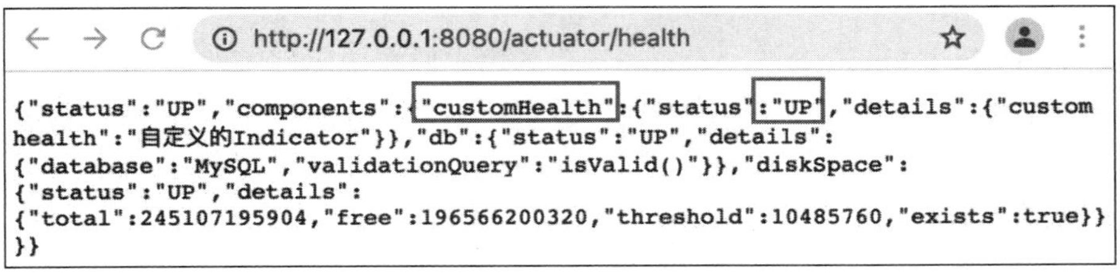

图 8.25　自定义 HealthIndicator

在上面的示例中，返回值使用了Status.UP，这是预定义的4个Status对象之一，此外还有UNKNOWN、DOWN和OUT_OF_SERVICE。

在返回的结果中，最前面的是所谓的系统状态，这个状态是根据所有的HealthIndicator状态生成的。Spring Boot Actuator还提供对结果Status排序的接口StatusAggregator，其默认实现对预定义的4个Status做了排序，默认的排序顺序是down、out-of-service、unknown、up。在所有顺序中，最靠前的Status作为系统状态。

Spring Boot Actuator支持对Status进行自定义，而自定义可以通过自定义接口StatusAggregator的实现类或者通过配置项management.endpoint.health.status.order来修改Status的排序。在修改顺序时也可以加入自定义的Status。

我们可以自定义StatusAggregator的实现类，或者给自定义的Status指定顺序。

【示例8.10】　自定义Status并指定Status的排列顺序

在配置文件application.yml中添加配置，指定Status的顺序，这里添加上自定义的status和custom，并指定custom的顺序在unknown之后、在up之前。具体配置代码如下：

```
management:
  endpoint:
    health:
      status:
        order: "down,out-of-service,unknown,custom,up"
```

然后修改CustomHealth的代码，自定义Status为CUSTOM，修改后代码如下：

```
return new Health.Builder(new Status("CUSTOM",""), map).build();
```

重启项目，在浏览器中访问Health Endpoint，这时系统Status已经成为CUSTOM，如图8.26所示。

图 8.26 自定义 Status 为 CUSTOM

最后，说明一下components中的排列，其顺序是按照key的字母顺序排列的。

8.2.5 Metrics Endpoint

Metrics Endpoint是Spring Boot Actuator提供的用来监控应用程序性能的模块。监控性能具体来讲包括两部分，一部分是性能数据的收集，另一部分是性能数据的存储和展示。

1. 性能数据的收集

性能数据的收集被称作Metrics，需要通过代码来完成。Spring Boot Actuator内置了18个Metrics，覆盖了多种技术。

这些Metrics分别是JVM Metrics、System Metrics、Logger Metrics、Spring MVC Metrics、Spring WebFlux Metrics、Jersey Server Metrics、HTTP Client Metrics、Tomcat Metrics、Cache Metrics、DataSource Metrics、Hibernate Metrics、Spring Data Repository Metrics、RabbitMQ Metrics、Spring Integration Metrics、Kafka Metrics、MongoDB Metrics和@Timed Annotation Support。

从Metrics的命名上大致可以看出相关的技术，这些Metrics大部分都会被默认配置，并在项目使用相关技术时自动开启。我们不需要在一开始就掌握所有的细节，了解大概有哪些功能即可，等需要使用时再查看具体信息。

2. 性能监控系统

市面上有许多成熟的性能监控系统，可以用来记录和展示性能数据。Spring Boot Metrics支持将Metrics收集的性能数据传递给性能监控执行，从而能够比较容易地实现应用从性能数据收集到展示得相对完整的性能监控链条。

Spring Boot Actuator默认使用Micrometer框架作为性能监控的门面技术，并支持对接18种监控系统，分别为Simple、AppOptics、Atlas、Datadog、Dynatrace、Elastic、Ganglia、Graphite、Humio、Influx、JMX、KairosDB、New Relic、Prometheus、SignalFx、Stackdriver、StatsD和Wavefront。

在这些监控系统中，Simple是Micrometer提供的内存型监控系统，也是Spring Boot Actuator默认启用的监控系统。通过Metrics Endpoint查看到的信息就是Simple提供的。当存在其他可用的性能监控系统时，会自动禁用Simple，也可以通过配置项management.metrics.export.simple.enabled=false来显式禁用Simple。

3. Metrics Endpoint 的使用

Metrics Endpoint默认没有开放访问，所以要使用Metrics Endpoint，需要先开放其访问功能。比如可以通过如下配置开放Metrics的HTTP访问：

```yaml
management:
  endpoints:
    web:
      exposure:
        include: "health,metrics"
```

Metrics Endpoint的默认访问路径是/actuator/metrics。在浏览器中访问http://127.0.0.1:8080/actuator/metrics，会以数组形式返回当前Metrics所有可用的Meter名称，响应结果如图8.27所示。

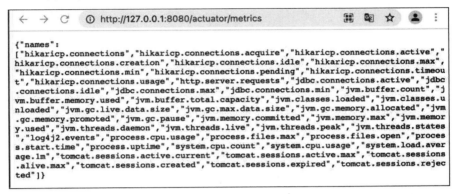

图 8.27　Metric Endpoint 支持的所有 Meter

所谓Meter，其实就是Metrics的进一步细分，是细粒度的指标。

将上面的响应结果格式化后会看得更清晰一些，格式化后的数据如下：

```
{
    "names":[
        "hikaricp.connections",
        "hikaricp.connections.acquire",
        "hikaricp.connections.active",
        "hikaricp.connections.creation",
        "hikaricp.connections.idle",
        "hikaricp.connections.max",
        "hikaricp.connections.min",
        "hikaricp.connections.pending",
        "hikaricp.connections.timeout",
        "hikaricp.connections.usage",
        "http.server.requests",
        "jdbc.connections.active",
        "jdbc.connections.idle",
        "jdbc.connections.max",
        "jdbc.connections.min",
        "jvm.buffer.count",
        "jvm.buffer.memory.used",
        "jvm.buffer.total.capacity",
        "jvm.classes.loaded",
        "jvm.classes.unloaded",
        "jvm.gc.live.data.size",
        "jvm.gc.max.data.size",
        "jvm.gc.memory.allocated",
```

```
            "jvm.gc.memory.promoted",
            "jvm.gc.pause",
            "jvm.memory.committed",
            "jvm.memory.max",
            "jvm.memory.used",
            "jvm.threads.daemon",
            "jvm.threads.live",
            "jvm.threads.peak",
            "jvm.threads.states",
            "log4j2.events",
            "process.cpu.usage",
            "process.files.max",
            "process.files.open",
            "process.start.time",
            "process.uptime",
            "system.cpu.count",
            "system.cpu.usage",
            "system.load.average.1m",
            "tomcat.sessions.active.current",
            "tomcat.sessions.active.max",
            "tomcat.sessions.alive.max",
            "tomcat.sessions.created",
            "tomcat.sessions.expired",
            "tomcat.sessions.rejected"
        ]
    }
```

这些Meter名称以"."分隔成了多个部分，从这些Meter名称的第一部分可以推断出示例中应用程序开启的Metrics。比如hikaricp和jdbc是DataSource Metrics的Meter的前缀，tomcat是Tomcat Metrics的Meter的前缀，process是System Metrics的Meter的前缀。在必要的时候，具体Meter命名前缀可以通过Spring Boo Actuator官方文档来查询。

对于具体Meter的访问，通过Metrics Endpoint的URL拼接上"/"，再拼接上Meter名称来访问。比如对于meterhikaricp.connections，默认条件下其访问地址就是/actuator/metrics/hikaricp.connections。在浏览器中访问地址http://127.0.0.1:8080/actuator/metrics/hikaricp.connections，响应结果如图8.28所示。

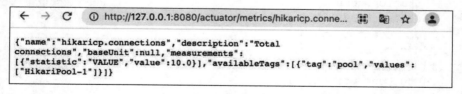

图 8.28　查看 hikaricp.connections

【示例8.11】　通过Metrics Endpoint查看JVM Metrics提供的信息

JVM Metrics是Spring Boot Actuator内置的Metrics之一，用来获取JVM性能相关的信息。这些信息有内存和缓存池的详细信息，垃圾收集器的统计信息，线程使用的信息和类的加载、卸载数量等。

JVM Metrics以"jvm."为Meter的前缀。从前面的Meter名称列表中可以看出，属于JVM的Meter又可以分成缓存（Buffer）、类加载（Classes）、垃圾回收（GC）、内存（Memory）和线程（Threads），基本涵盖了关于JVM的大部分指标。

比如现在要查看当前可用的内存，使用meter "jvm.memory.committed"，所以在浏览器中访问地址http://127.0.0.1:8080/actuator/metrics/jvm.memory.committed，响应结果如图8.29所示。

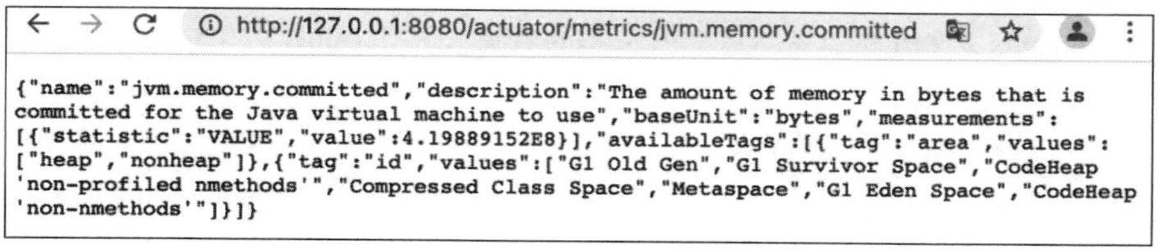

图 8.29　查看 JVM 的可用内存

对响应结果格式化，格式化后的数据如下：

```
{
    "name":"jvm.memory.committed",
    "description":"The amount of memory in bytes that is committed for the Java virtual machine to use",
    "baseUnit":"bytes",
    "measurements":[
        {
            "statistic":"VALUE",
            "value":419889152
        }
    ],
    "availableTags":[
        {
            "tag":"area",
            "values":[
                "heap",
                "nonheap"
            ]
        },
        {
            "tag":"id",
            "values":[
                "G1 Old Gen",
                "G1 Survivor Space",
                "CodeHeap 'non-profiled nmethods'",
                "Compressed Class Space",
                "Metaspace",
                "G1 Eden Space",
                "CodeHeap 'non-nmethods'"
            ]
        }
    ]
}
```

其中：

- name 就是当前 Meter 的名称。
- description 是对当前的描述，直译过来就是"提交给 Java 虚拟机使用的内存量（以字节为单位）"。
- baseUnit 是值的单位，这里说明了是 bytes，也就是字节。
- measurements 是测量的结果，也是最关心的数据。这里 value 显示为 419889152，单位是字节，大概是 400MB 左右。
- availableTags 说明当前 Meter 还有更细粒度的信息可以查看，通过在已有的 URL 上拼接参数来使用。其拼接语法为 "?tag=<tag>:<value>"，有多个 tag 时通过&符号拼接。

这里有两个tag可选，分别是area和id。比如可以指定tag为area，值为heap，那么URL就是/actuator/metrics/jvm.memory.committed?tag=area:heap。在浏览器中访问这个地址，响应如图8.30所示。

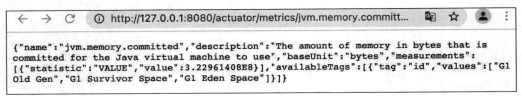

图 8.30　查看 JVM heap 的可用内存

对响应结果格式化，格式化之后的数据如下：

```
{
    "name":"jvm.memory.committed",
    "description":"The amount of memory in bytes that is committed for the Java virtual machine to use",
    "baseUnit":"bytes",
    "measurements":[
        {
            "statistic":"VALUE",
            "value":322961408
        }
    ],
    "availableTags":[
        {
            "tag":"id",
            "values":[
                "G1 Old Gen",
                "G1 Survivor Space",
                "G1 Eden Space"
            ]
        }
    ]
}
```

数据结构与没有指定tag时类似，但是具体数据有差别。比如这里的统计值变成了322961408，大概300MB左右。结合tag为heap，这里值的含义就是heap可用内存为300MB左右。

再看最后一个字段availableTags，因为当前指定了tag为heap，所以在此基础上可以指定的tag只有id，并且可选的值也只剩下3个，因为只有这3个id是属于堆的。可以在原有地址上拼接上?tag=id:G1 Old Gen，其含义是只查看G1的老年代区域。在浏览器中访问地址http://127.0.0.1:8080/actuator/metrics/jvm.memory.committed?tag=area:heap&tag=id:G1%20Old%20Gen，响应结果如图8.31所示。

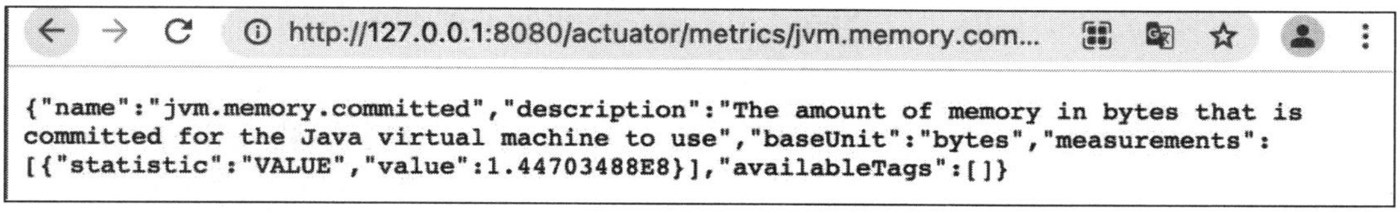

```
{"name":"jvm.memory.committed","description":"The amount of memory in bytes that is
committed for the Java virtual machine to use","baseUnit":"bytes","measurements":
[{"statistic":"VALUE","value":1.44703488E8}],"availableTags":[]}
```

图 8.31　查看 JVM G1 老年代的可用内存

对响应结果格式化，格式化后的数据如下：

```
{
    "name":"jvm.memory.committed",
    "description":"The amount of memory in bytes that is committed for the Java
virtual machine to use",
    "baseUnit":"bytes",
    "measurements":[
        {
            "statistic":"VALUE",
            "value":144703488
        }
    ],
    "availableTags":[

    ]
}
```

可以看到统计信息变成144703488，大概138MB，说明当前JVM堆的老年代可用内存大概为138MB。同时availableTags没有可选项，也就是说，当前查询的数据已经是所提供的最细粒度。通过由粗到细，可以一步步分析出当前JVM的内存使用情况。

8.2.6　自定义 Endpoint

Spring Boot Actuator支持自定义Endpoint，并为此提供一系列的注解。本节将介绍如何通过注解来自定义Endpoint。

1. 标注@Bean 的 Endpoint 注解

类作为自定义Endpoint的单元，Spring Boot提供了多个用户标注类为Endpoint的注解。这些注解分别有@Endpoint、@JmxEndpoint、@WebEndpoint、@ServletEndpoint、@ControllerEndpoint和@RestControllerEndpoint。

@Endpoint、@JmxEndpoint、@WebEndpoint注解的作用类似，功能上稍有区别。@Endpoint是不区分开放访问的技术，而@JmxEndpoint和@WebEndpoint只会对各自支持的技术开放。

@ServletEndpoint注解可以用来标注Servlet实现类的父类，以使Servlet实现类成为Endpoint类。这种定义Endpoint的方式好处是可以使用更多的Servlet特性，不过缺点是损失了Endpoint的可移植性。这种方式比较适合已经存在的Servlet，而想做的只是将已有的Servlet作为Endpoint开放。对于新开发的Endpoint，并不推荐使用这种方式。

最后的两个注解@ControllerEndpoint和@RestControllerEndpoint用来标注Endpoint类时，在Endpoint类内部定义的方法可以使用Spring MVC或者Spring WebFlux的注解来标注，以此来开放。这种方式同样是可移植性差，对于新创建的Endpoint并不推荐。

以上注解都需要与定义为Bean的注解配合使用，单独使用上面的注解并不会创建类实例。

这些注解使用上类似，都有两个属性：id和enableByDefault。id的值会作为自定义Endpoint的id。enableByDefault配置当前Endpoint是否默认启用，该属性默认值为true。

要开放自定义的Endpoint，如果配置文件中配置了配置项management.endpoints.web.exposure.include，那么需要在该配置项中添加上自定义Endpoint的id。

2. Web Endpoint 的方法定义

当类被标注为Endpoint时，如果采用了HTTP的开放方式，那么其默认访问地址就是"/actuator/"拼接上自定义的id。Spring Boot还提供了注解，用于将HTTP的method和类实例的方法绑定。

这些用于和HTTP method 绑定的注解有@ReadOperation、@WriteOperation和@DeleteOperation，分别绑定到GET、POST和DELETE。

被注解标记的方法的形参会被当作请求参数来处理，默认请求参数值不能为空。如果参数可选，可以通过注解@javax.annotation.Nullable或@org.springframework.lang.Nullable来标注可选参数对应的方法形参。

通过HTTP访问Endpoint时，支持在URL上拼接参数和在请求体中通过JSON格式封装参数，对于基本类型和String类型，框架会进行自动转换。

对于请求和响应的类型，@ReadOperation、@WriteOperation和@DeleteOperation注解与Spring MVC的@RequestMapping注解有一定的相似之处，比如它们都有用来执行返回值类型的属性produces，可以通过此属性指定响应头的Content-Type。

与注解@RequestMapping不同的是，这3个注解没有属性consumes，也就是不能自定义请求的内容格式。在这3个注解中，只有@WriteOperation隐含consumes为application/vnd.spring-boot.actuator.v2+json, application/json。

3. Web Endpoint 可选的 Security 相关形参

对于通过HTTP开放的Endpoint，还有两个可以通过形参获取用户认证授权信息的类型，它们是java.security.Principal和org.springframework.boot.actuate.endpoint.SecurityContext。

从java.security.Principal类型的形参中可以获取当前用户的信息，从org.springframework.

boot.actuate.endpoint.SecurityContext类型的形参中可以获取用户角色的信息。

【示例8.12】 自定义HTTP Endpoint

新创建一个类，命名为CustomEndpoint。使用注解@Component标注，实例交由Spring容器管理。再使用注解@Endpoint标注，声明为Endpoint，并指定id为customend。

在类中创建方法customEndpoint，接收String类型的参数param，并且使用注解@ReadOperation标注方法，从而使得方法和HTTP Get绑定。方法返回String类型的返回值，以供页面展示。具体代码如下：

```
package com.example.demo.endpoint;

import org.springframework.boot.actuate.endpoint.annotation.Endpoint;
import org.springframework.boot.actuate.endpoint.annotation.ReadOperation;
import org.springframework.stereotype.Component;

@Component
@Endpoint(id="customend")
public class CustomEndpoint {
    @ReadOperation
    public String customEndpoint(String param){
        return "自定义的Endpoint，参数为"+param;
    }
}
```

在配置文件中修改配置，开放自定义的id为customend的Endpoint，配置代码如下：

```
management:
  endpoints:
    web:
      exposure:
        include: "health,customend"
```

至此，便完成了一个自定义的Endpoint。在浏览器中访问Actuator发现页，即http://127.0.0.1:8080/actuator，可以看到自定义的Endpoint及其地址，如图8.32所示。

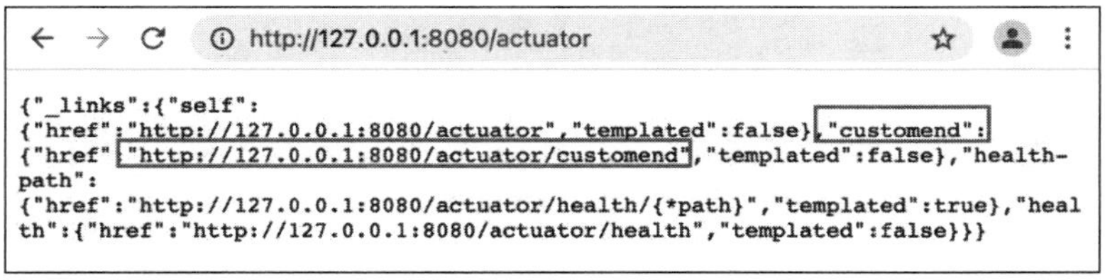

图8.32 自定义Endpoint之后的Actuator发现页

直接访问customend的地址http://127.0.0.1:8080/actuator/customend，响应结果如图8.33所示。

响应中提示Bad Request和400，这是因为在customEndpoint方法的形参列表中定义了一个参数param，而参数默认是必选参数，所以访问时必须传递参数。这样在URL后拼接参数"?param=--测试自定义的endpoint--"，即在浏览器中访问"http://127.0.0.1:8080/actuator/customend?param=--测试自定义的endpoint--"，响应如图8.34所示。

图 8.33　没有传递自定义的参数

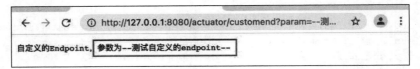

图 8.34　自定义 Endpoint 的响应结果

方法成功返回，并且正确获取到了参数信息。

第 9 章

Spring Boot数据访问

对数据访问首先要连接上数据库，以及数据库中定义项目运行所必需的元数据。前面章节介绍了一些操作数据的框架，本章将介绍Spring Boot支持的数据源管理和数据元数据管理的相关技术和使用。

本章主要涉及的知识点有：

⌘ Spring Boot如何自动配置默认数据源。
⌘ 如何在项目中配置多个数据源。
⌘ 如何初始化数据库。

9.1 自动配置默认数据源

自动配置是Spring Boot的一大特性，本节我们来看一下Spring Boot是如何配置默认数据源的。

Spring Boot的自动配置类都在spring-boot-autoconfigure-2.5.3.jar这个依赖下，关于数据源DataSource的配置可以定位到org.springframework.boot.autoconfigure.jdbc这个包的DataSourceAutoConfiguration类下。

DataSourceAutoConfiguration类是一个配置类，其部分代码如下：

```
@Configuration(proxyBeanMethods = false)
@ConditionalOnClass({ DataSource.class, EmbeddedDatabaseType.class })
@ConditionalOnMissingBean(type = "io.r2dbc.spi.ConnectionFactory")
@EnableConfigurationProperties(DataSourceProperties.class)
@Import({ DataSourcePoolMetadataProvidersConfiguration.class,
    DataSourceInitializationConfiguration.
InitializationSpecificCredentialsDataSourceInitializationConfiguration.class,
    DataSourceInitializationConfiguration.
SharedCredentialsDataSourceInitializationConfiguration.class })
    public class DataSourceAutoConfiguration {
```

从代码中可以看到配置生效，类路径下必须有两个类：DataSource.class 和 EmbeddedDatabaseType.class。这里的DataSource是javax.sql包下的，是JDK提供的，所以一般来说这个类都是存在的。第二个类EmbeddedDatabaseType是一个枚举类，位于org.springframework.jdbc.datasource.embedded包下，是spring-jdbc.jar提供的。所以通常来说，只要项目直接或间接地引用了spring-jdbc，便会开启DataSource的自动配置。

【示例9.1】 使用Spring Data JPA间接依赖spring-jdbc

在使用Spring Data JPA时，只需要引入spring-data-starter-jpa，然后会间接依赖到spring-jdbc，具体依赖路径如图9.1所示。

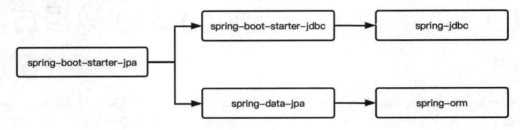

图 9.1 使用 JPA 时 spring-jdbc 的依赖路径

【示例9.2】 使用MyBatis间接依赖spring-jdbc

使用MyBatis时也是通过MyBatis提供的Starter来添加依赖的，同样会间接依赖spring-jdbc，只不过依赖路径不同，具体如图9.2所示。

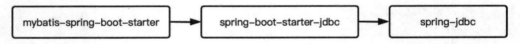

图 9.2 使用 MyBatis 时 spring-jdbc 的依赖路径

以上是第一个条件。接下来看第二个条件，条件代码如下：

```
@ConditionalOnMissingBean(type = "io.r2dbc.spi.ConnectionFactory")
```

这里要求容器内没有r2dbc的ConnectionFactory实例，r2dbc是响应式访问数据库的技术，如果没有应用响应式，也不会实例化此类实例，所以第二个条件对于非应用响应式技术栈的项目通常也是符合的。

接下来便在容器中注入DataSource的配置类，代码如下：

```
@EnableConfigurationProperties(DataSourceProperties.class)
```

这里会实例化DataSourceProperties类。DataSourceProperties类位于org.springframework.boot.autoconfigure.jdbc包下，也是自动配置的一部分，类中的部分代码如下：

```
@ConfigurationProperties(prefix = "spring.datasource")
public class DataSourceProperties implements BeanClassLoaderAware,
InitializingBean {
    private ClassLoader classLoader;
    // ......
```

```java
    private String name;
    private Class<? extends DataSource> type;
    private String driverClassName;
    private String url;
    private String username;
    private String password;
    // ......
     private List<String> schema;
    // ......
    private String separator = ";";
    // ......
}
```

这个类定义中指定了对应配置项的前缀spring.datasource，这个类中定义的字段如url、username、password等，也是我们在使用自动配置数据源时常用的配置项。

在加载了配置之后，在类中如下代码：

```java
@Configuration(proxyBeanMethods = false)
@Conditional(PooledDataSourceCondition.class)
@ConditionalOnMissingBean({ DataSource.class, XADataSource.class })
@Import({ DataSourceConfiguration.Hikari.class, DataSourceConfiguration.Tomcat.class,
        DataSourceConfiguration.Dbcp2.class, DataSourceConfiguration.OracleUcp.class,
        DataSourceConfiguration.Generic.class, DataSourceJmxConfiguration.class })
protected static class PooledDataSourceConfiguration {

}
```

导入了DataSourceConfiguration类中对数据源的配置,可以看到此处引入了Hikari、Tomcat、Dbcp2、OracleUcp和Generic这5类数据源。创建这5类数据源的实例代码也在DataSourceConfiguration中，具体代码如下：

```java
@Configuration(proxyBeanMethods = false)
@ConditionalOnClass(org.apache.tomcat.jdbc.pool.DataSource.class)
@ConditionalOnMissingBean(DataSource.class)
@ConditionalOnProperty(name = "spring.datasource.type", havingValue = "org.apache.tomcat.jdbc.pool.DataSource",
        matchIfMissing = true)
static class Tomcat {

    @Bean
    @ConfigurationProperties(prefix = "spring.datasource.tomcat")
    org.apache.tomcat.jdbc.pool.DataSource dataSource(DataSourceProperties properties) {
        org.apache.tomcat.jdbc.pool.DataSource dataSource = createDataSource
                (properties,org.apache.tomcat.jdbc.pool.DataSource.class);
        DatabaseDriver databaseDriver = DatabaseDriver.fromJdbcUrl
                (properties.determineUrl());
        String validationQuery = databaseDriver.getValidationQuery();
```

```java
            if (validationQuery != null) {
                dataSource.setTestOnBorrow(true);
                dataSource.setValidationQuery(validationQuery);
            }
            return dataSource;
        }

    }

    @Configuration(proxyBeanMethods = false)
    @ConditionalOnClass(HikariDataSource.class)
    @ConditionalOnMissingBean(DataSource.class)
    @ConditionalOnProperty(name = "spring.datasource.type", havingValue =
"com.zaxxer.hikari.HikariDataSource",
            matchIfMissing = true)
    static class Hikari {

        @Bean
        @ConfigurationProperties(prefix = "spring.datasource.hikari")
        HikariDataSource dataSource(DataSourceProperties properties) {
            HikariDataSource dataSource = createDataSource(properties,
HikariDataSource.class);
            if (StringUtils.hasText(properties.getName())) {
                dataSource.setPoolName(properties.getName());
            }
            return dataSource;
        }

    }

    @Configuration(proxyBeanMethods = false)
    @ConditionalOnClass(org.apache.commons.dbcp2.BasicDataSource.class)
    @ConditionalOnMissingBean(DataSource.class)
    @ConditionalOnProperty(name = "spring.datasource.type", havingValue =
"org.apache.commons.dbcp2.BasicDataSource",
            matchIfMissing = true)
    static class Dbcp2 {

        @Bean
        @ConfigurationProperties(prefix = "spring.datasource.dbcp2")
        org.apache.commons.dbcp2.BasicDataSource dataSource(DataSourceProperties
properties) {
            return createDataSource(properties, org.apache.commons.dbcp2.
BasicDataSource.class);
        }

    }

    @Configuration(proxyBeanMethods = false)
    @ConditionalOnClass({ PoolDataSourceImpl.class, OracleConnection.class })
    @ConditionalOnMissingBean(DataSource.class)
    @ConditionalOnProperty(name = "spring.datasource.type", havingValue =
            "oracle.ucp.jdbc.PoolDataSource", matchIfMissing = true)
    static class OracleUcp {
```

```
    @Bean
    @ConfigurationProperties(prefix = "spring.datasource.oracleucp")
    PoolDataSourceImpl dataSource(DataSourceProperties properties) throws
SQLException {
        PoolDataSourceImpl dataSource = createDataSource(properties,
PoolDataSourceImpl.class);
        dataSource.setValidateConnectionOnBorrow(true);
        if (StringUtils.hasText(properties.getName())) {
            dataSource.setConnectionPoolName(properties.getName());
        }
        return dataSource;
    }
}

@Configuration(proxyBeanMethods = false)
@ConditionalOnMissingBean(DataSource.class)
@ConditionalOnProperty(name = "spring.datasource.type")
static class Generic {
    @Bean
    DataSource dataSource(DataSourceProperties properties) {
        return properties.initializeDataSourceBuilder().build();
    }
}
```

每一类数据源都使用了@ConditionalOnMissingBean(DataSource.class)注解来标注，其结果是这5类数据源在自动配置中即使有多个数据源符合实例化条件，也只会实例化其中的一个。

这5类数据源是通过@Import导入的，其顺序会被当作加载顺序，最后一个数据源Generic的条件十分宽松，只需要容器中没有DataSource的实例，所以当前4个数据源没有创建时便会创建DataSource数据源。

@Import指定的第一个数据源HikariDataSource，也是Spring Boot默认支持的数据源。可以看到，这个数据源需要类路径下存在HikariDataSource.class，而HikariDataSource在HikariCP-4.0.3.jar包下。又有spring-boot-starter-jdbc中依赖了HikariCP，所以Spring Boot默认会创建HikariDataSource实例作为数据源。

在上面实例化5类数据源的代码中，除了Generic外，其他4类数据源都是通过createDataSource方法来实例化各自的数据源类，createDataSource方法的代码如下：

```
    protected static <T> T createDataSource(DataSourceProperties properties,
Class<? extends DataSource> type) {
        return (T) properties.initializeDataSourceBuilder().type(type).build();
    }
```

可以看到createDataSource方法，甚至Generic数据源的初始化都使用DataSourceProperties实例来实例化DataSource的实现类。至此，便完成了默认数据源的实例化。

9.2 自定义一个或多个数据源

当自动配置的数据源不能满足对数据源配置的需要，或者需要配置多个数据源时，我们需要自定义数据源。

自定义数据源可以通过在配置类中使用@Bean注解来定义，比如通过以下代码定义数据源CustomDataSource：

```
@Configuration(proxyBeanMethods = false)
public class CustomDataSourceConfiguration {

    @Bean
    @ConfigurationProperties(prefix = "custom.datasource")
    public SomeOneDataSource dataSource() {
        return new SomeOneDataSource();
    }

}
```

这里的SomeOneDataSource是DataSource接口的实现类，可以是自定义实现的，也可以是其他技术提供的实现类。

上面的dataSource()方法之上还标注了@ConfigurationProperties注解，并指定前缀为custom.datasource。从而Spring Environment中定义的以custom.datasource开头的配置项会赋值给SomeOneDataSource实例的同名属性。

9.2.1 在使用默认数据源实例的基础上自定义配置

前面的代码通过new关键字来创建DataSource实例，如果只是想在Spring Boot自定义数据源的基础上丰富配置，那么大可不必通过new关键字来创建实例。Spring Boot提供了使用默认数据源实例化的API，参考如下代码：

```
@Bean
@ConfigurationProperties("custom.datasource")
public DataSource dataSource() {
    return DataSourceBuilder.create().build();
}
```

这里会返回Spring Boot匹配到的数据源实例，同时可以通过配置项custom.datasource来为数据源实例配置属性。不过这么一来，缺点就是无法使用Spring Boot提供的对配置项的自动转换，比如，HikariDataSource的数据库地址属性名为jdbcUrl，从而在配置项中应该命名为jdbc-url，而不能是url。

也可以使用DataSourceProperties来实现配置项的转换，并结合DataSourceProperties的实例方法initializeDataSourceBuilder()来创建数据源，可以参考如下代码：

```
@Configuration(proxyBeanMethods = false)
public class CustomDataSource {
    @Bean
    @Primary
    @ConfigurationProperties("custom.datasource")
    public DataSourceProperties dataSourceProperties() {
        return new DataSourceProperties();
    }
    @Bean
    public DataSource dataSource(DataSourceProperties dataSourceProperties){
        return dataSourceProperties.initializeDataSourceBuilder().build();
    }
}
```

对应的application.yaml中的配置如下：

```
custom:
  datasource:
    url: "jdbc:mysql://127.0.0.1:3306/demo" #这里是作者 MySQL 实例的地址
    username: "root" #
    password: "123456" #
```

如果有配置项是DataSourceProperties不支持的，那么可以在dataSource方法上使用注解@ConfigurationProperties来指定配置项的前缀。

另外，DataSourceBuilder还提供了指定数据库类型的方法type，通过该方法的参数执行生成DataSource的具体类型，比如上面的示例代码中指定为HikariDataSource，那么代码可以修改如下：

```
@Configuration(proxyBeanMethods = false)
public class CustomDataSource {
    @Bean
    @Primary
    @ConfigurationProperties("custom.datasource")
    public DataSourceProperties dataSourceProperties() {
        return new DataSourceProperties();
    }
    @Bean
    public HikariDataSource dataSource(DataSourceProperties dataSourceProperties){
        return dataSourceProperties.initializeDataSourceBuilder().
type(HikariDataSource.class).build();
    }
}
```

9.2.2 配置多个数据源

配置多个数据源指的是创建多个DataSource实例，这些实例可以是同一种技术实现的，也可以是不同的技术实现的。比如，可以两个数据源分别对应两个数据库，数据库名为demo和demo1。application.yaml配置文件中的代码如下：

```yaml
my:
  datasource1:
    url: "jdbc:mysql://127.0.0.1:3306/demo1"
    username: "root"
    password: "123456"
  datasource2:
    url: "jdbc:mysql://127.0.0.1:3306/demo2"
    username: "root"
    password: "123456"
```

在配置类中创建DataSource实例，可以通过DataSourceProperties类来创建，配置类代码如下：

```java
import com.zaxxer.hikari.HikariDataSource;
import org.springframework.boot.autoconfigure.jdbc.DataSourceProperties;
import org.springframework.boot.context.properties.ConfigurationProperties;
import org.springframework.context.annotation.Bean;
import org.springframework.context.annotation.Configuration;
import org.springframework.context.annotation.Primary;

@Configuration
public class DataSourceConfig {
    @Bean
    @ConfigurationProperties("my.datasource1")
    public DataSourceProperties dataSourceProperties1(){
        return new DataSourceProperties();
    }

    @Bean
    @Primary
    public HikariDataSource dataSource1(DataSourceProperties dataSourceProperties1){
        return dataSourceProperties1.initializeDataSourceBuilder().type(HikariDataSource.class).build();
    }

    @Bean
    @ConfigurationProperties("my.datasource2")
    public DataSourceProperties dataSourceProperties2(){
        return new DataSourceProperties();
    }

    @Bean
    public HikariDataSource dataSource2(DataSourceProperties dataSourceProperties2){
        return dataSourceProperties2.initializeDataSourceBuilder().type(HikariDataSource.class).build();
    }
}
```

这里注意，在创建多个数据源时，需要使用@Primary指定一个数据源作为默认数据源，这样才能使容器初始化的后续工作正常进行，上面的代码将@Primary标注到了数据源dataSource1上。在测试中，不使用注解@Primary标注数据源，或者将上面的两个数据源都标注上@Primary，项目启动没有报错，但是没有执行数据库连接池的初始化和JPA EntityManagerFactory的初始化。

9.3　Spring Data JPA与数据源绑定

当配置了多个数据源的时候，需要设置框架使用的是配置源中的哪个。对于不同的持久层框架，与数据源绑定的方式也不同，比如MyBatis可以通过sessionFactory或者动态数据源的技术来实现。本节将介绍Spring Data JPA如何使用多个数据源。

JPA绑定多个数据源时，需要为每个数据源创建EntityManager实例以及TransactionManager实例。创建EntityManager实例可以借助Spring ORM提供的LocalContainerEntityManagerFactoryBean类来实现。当项目依赖spring-data-jpa时，会间接依赖spring-orm，依赖路径如图9.3所示。

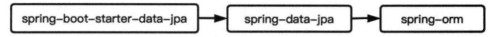

图9.3　JPA 到 spring-orm 的依赖路径

在创建EntityManager时，需要指定数据源扫描Repository所在的路径和实体类所在的路径，先创建两个实体类Customer和Manager，以及它们对应的Repository接口CustomerRepository和ManagerRepository，这4个类所属路径如图9.4所示。

图9.4　项目目录接口

实体类Customer代码如下：

```
package com.demo.testds.ds1.entity;

import javax.persistence.*;

@Entity
@Table(name = "customers")
public class Customer {
    @Id
    @GeneratedValue(strategy = GenerationType.IDENTITY)
    private Integer id;
    private String customerName;
}
```

实体类Manager代码如下：

```
package com.demo.testds.ds2.entity;

import javax.persistence.*;

@Entity
@Table(name = "managers")
public class Manager {
    @Id
    @GeneratedValue(strategy = GenerationType.IDENTITY)
    private Integer id;
    private String managerName;
```

}

CustomerRepository接口和ManagerRepository接口都继承了JpaRepository、JpaSpecificationExecutor两个接口，没有额外定义方法。

对于数据源，使用上一节最后的示例代码定义的两个数据源实例dataSource1和dataSource2，它们分别访问数据库demo1和demo2。接下来需要创建两个EntityManager实例，使得Customer和CustomerRepository使用数据库demo1，Manager和ManagerRepository使用数据库demo2。

为了增加代码的可读性，将两个EntityManager实例的创建放在两个类中来配置。

首先实现数据源dataSource1的EntityManager实例的创建。在config目录下创建类DS1EntityManagerFactoryConfig，具体代码如下：

```java
package com.demo.testds.config;

import org.springframework.beans.factory.annotation.Autowired;
import org.springframework.beans.factory.annotation.Qualifier;
import org.springframework.boot.autoconfigure.orm.jpa.HibernateProperties;
import org.springframework.boot.autoconfigure.orm.jpa.HibernateSettings;
import org.springframework.boot.autoconfigure.orm.jpa.JpaProperties;
import org.springframework.boot.orm.jpa.EntityManagerFactoryBuilder;
import org.springframework.context.annotation.Bean;
import org.springframework.context.annotation.Configuration;
import org.springframework.context.annotation.Primary;
import org.springframework.data.jpa.repository.config.EnableJpaRepositories;
import org.springframework.orm.jpa.JpaTransactionManager;
import org.springframework.orm.jpa.LocalContainerEntityManagerFactoryBean;
import org.springframework.transaction.PlatformTransactionManager;
import org.springframework.transaction.annotation.EnableTransactionManagement;
import javax.persistence.EntityManager;
import javax.sql.DataSource;

@Configuration
@EnableTransactionManagement
@EnableJpaRepositories(
        entityManagerFactoryRef = "ds1EntityManagerFactory",
        transactionManagerRef = "ds1TransactionManager",
        basePackages = {"com.demo.ds.ds1.dao"})
public class DS1EntityManagerFactoryConfig {

    @Autowired
    @Qualifier("dataSource1")
    private DataSource dataSource1;

    @Autowired
    private HibernateProperties hibernateProperties;

    @Bean(name = "ds1EntityManager")
    public EntityManager ds1EntityManager(EntityManagerFactoryBuilder builder, JpaProperties jpaProperties) {
        return ds1EntityManagerFactory(builder, jpaProperties).getObject().
```

```
createEntityManager();
        }

        @Bean(name = "ds1EntityManagerFactory")
        @Primary
        public LocalContainerEntityManagerFactoryBean ds1EntityManagerFactory
(EntityManagerFactoryBuilder builder, JpaProperties jpa1Properties) {
            return builder
                    .dataSource(dataSource1)
                    .properties(hibernateProperties.determineHibernateProperties(
jpa1Properties.getProperties(), new
                            HibernateSettings()))
                    .packages("com.demo.testds.ds1.entity")
                    .persistenceUnit("ds1PersistenceUnit")
                    .build();
        }

        @Bean
        PlatformTransactionManager ds1TransactionManager
(EntityManagerFactoryBuilder builder, JpaProperties jpaProperties) {
            return new JpaTransactionManager(ds1EntityManagerFactory(builder,
jpaProperties).getObject());
        }
    }
```

然后可以依照配置类DS1EntityManagerFactoryConfig的实现完成数据源dataSource2的EntityManager的实例化配置。创建配置类DS2EntityManagerFactoryConfig，具体代码如下：

```
package com.demo.testds.config;

import org.springframework.beans.factory.annotation.Autowired;
import org.springframework.beans.factory.annotation.Qualifier;
import org.springframework.boot.autoconfigure.orm.jpa.HibernateProperties;
import org.springframework.boot.autoconfigure.orm.jpa.HibernateSettings;
import org.springframework.boot.autoconfigure.orm.jpa.JpaProperties;
import org.springframework.boot.orm.jpa.EntityManagerFactoryBuilder;
import org.springframework.context.annotation.Bean;
import org.springframework.context.annotation.Configuration;
import org.springframework.data.jpa.repository.config.
EnableJpaRepositories;
import org.springframework.orm.jpa.JpaTransactionManager;
import org.springframework.orm.jpa.LocalContainerEntityManagerFactoryBean;
import org.springframework.transaction.PlatformTransactionManager;
import org.springframework.transaction.annotation.
EnableTransactionManagement;

import javax.persistence.EntityManager;
import javax.sql.DataSource;

@Configuration
@EnableTransactionManagement
@EnableJpaRepositories(
        entityManagerFactoryRef = "ds2EntityManagerFactory",
```

```java
        transactionManagerRef = "ds2TransactionManager",
        basePackages = {"com.demo.ds.ds2.dao"})
public class DS2EntityManagerFactoryConfig {
    @Autowired
    JpaProperties jpaProperties;

    @Autowired
    @Qualifier("dataSource2")
    DataSource dataSource2;

    @Autowired
    private HibernateProperties hibernateProperties;

    @Bean(name = "ds2EntityManager")
    public EntityManager ds2EntityManager(EntityManagerFactoryBuilder builder, JpaProperties jpaProperties) {
        return ds2EntityManagerFactory(builder, jpaProperties).getObject().createEntityManager();
    }

    @Bean(name = "ds2EntityManagerFactory")
    public LocalContainerEntityManagerFactoryBean ds2EntityManagerFactory(EntityManagerFactoryBuilder builder, JpaProperties jpaProperties) {
        return builder
                .dataSource(dataSource2)
                .properties(hibernateProperties.determineHibernateProperties(jpaProperties.getProperties(), new HibernateSettings()))
                .packages("com.demo.testds.ds2.entity")
                .persistenceUnit("ds2PersistenceUnit")
                .build();
    }

    @Bean
    PlatformTransactionManager ds2TransactionManager(EntityManagerFactoryBuilder builder, JpaProperties jpaProperties) {
        return new JpaTransactionManager(ds2EntityManagerFactory(builder, jpaProperties).getObject());
    }
}
```

新创建的两个配置类中都使用到了JpaProperties，对应配置文件application.yaml中spring.jpa配置项，所以在application.yaml中添加如下代码：

```yaml
spring:
  jpa:
    hibernate:
      ddl-auto: "update"
    show-sql: "true"
```

配置文件中指定了配置项spring.jpa.ddl-auto为update，所以，如果配置生效，那么在项目启动成功之后，应该会在数据库demo1和demo2中分别创建表customers和managers。

如果启动项目,可以在控制台看到创建数据源和初始化EntityManagerFactory的日志输出,如图9.5所示。

```
HHH000204: Processing PersistenceUnitInfo [name: ds1PersistenceUnit]
HHH000412: Hibernate ORM core version 5.4.32.Final
HCANN000001: Hibernate Commons Annotations {5.1.2.Final}
HikariPool-1 - Starting...
HikariPool-1 - Start completed.
HHH000400: Using dialect: org.hibernate.dialect.MySQL8Dialect
primary key (id)) engine=InnoDB
HHH000490: Using JtaPlatform implementation: [org.hibernate.engine.transaction.jta.platform.internal.NoJtaPlatform]
Initialized JPA EntityManagerFactory for persistence unit 'ds1PersistenceUnit'
HHH000204: Processing PersistenceUnitInfo [name: ds2PersistenceUnit]
HikariPool-2 - Starting...
HikariPool-2 - Start completed.
HHH000400: Using dialect: org.hibernate.dialect.MySQL8Dialect
primary key (id)) engine=InnoDB
HHH000490: Using JtaPlatform implementation: [org.hibernate.engine.transaction.jta.platform.internal.NoJtaPlatform]
Initialized JPA EntityManagerFactory for persistence unit 'ds2PersistenceUnit'
```

图 9.5　项目启动时的日志输出

在日志中可以看到两条建表语句,如图9.6所示。

```
2021-08-23 00:07:27.657  INFO 3999 --- [   main] org.hibernate.dialect.Dialect            : HHH000400: Using dialect: org.hibe
Hibernate: create table customers (id integer not null auto_increment, customer_name varchar(255), primary key (id)) engine=InnoDB
2021-08-23 00:07:28.454  INFO 3999 --- [   main] o.h.e.t.j.p.i.JtaPlatformInitiator       : HHH000490: Using JtaPlatform imple
2021-08-23 00:07:28.464  INFO 3999 --- [   main] j.LocalContainerEntityManagerFactoryBean : Initialized JPA EntityManagerFacto
2021-08-23 00:07:28.566  INFO 3999 --- [   main] o.hibernate.jpa.internal.util.LogHelper  : HHH000204: Processing Persistence
2021-08-23 00:07:28.573  INFO 3999 --- [   main] com.zaxxer.hikari.HikariDataSource       : HikariPool-2 - Starting...
2021-08-23 00:07:28.603  INFO 3999 --- [   main] com.zaxxer.hikari.HikariDataSource       : HikariPool-2 - Start completed.
2021-08-23 00:07:28.604  INFO 3999 --- [   main] org.hibernate.dialect.Dialect            : HHH000400: Using dialect: org.hibe
Hibernate: create table managers (id integer not null auto_increment, manager_name varchar(255), primary key (id)) engine=InnoDB
2021-08-23 00:07:28.648  INFO 3999 --- [   main] o.h.e.t.j.p.i.JtaPlatformInitiator       : HHH000490: Using JtaPlatform imple
```

图 9.6　DDL 语句日志输出

最后到数据库中查看,在数据库demo1中可以看到customers表,以及在实体类Customer中定义的字段信息,如图9.7所示。

在数据库demo2中可以看到managers表,以及在实体类Manager中定义的字段信息,如图9.8所示。

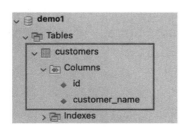
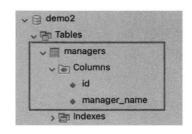

图 9.7　数据库 demo1 中生成的 customers 表　　图 9.8　数据库 demo2 中生成的 managers 表

9.4　数据库的初始化

这里数据库的初始化指的是对数据库中的元数据(表结构、索引信息等)进行创建或更新的过程。对数据库的管理不借助外部技术,以往都是手工来维护脚本文件的,在创建新环境或者对表结构进行更新时,手动执行所维护的脚本,以实现对数据库元数据的维护。原始的方式效率低，而且容易出错。鉴于此，Spring Boot提供了多种初始化数据库的方式，本节将介绍这些初始化数据库的方式。

9.4.1 基于 SQL 脚本初始化数据库

Spring Boot提供了基于SQL脚本初始化数据库的功能。默认情况下，此功能仅对内存型数据库有效，可以通过配置项spring.sql.init.mode来设置生效模式。比如，设置代码如下：

```
spring:
  sql:
    init:
      mode: always
```

使得对所有数据库类型都生效。此配置项的默认值是embedded，其含义就是对内置的内存型数据库生效。还可以配置如下：

```
spring:
  sql:
    init:
      mode: never
```

即指定配置值为never来关闭此功能。

使用SQL脚本初始化数据库可以分开支持DDL（Data Definition Language）和DML（Data Manipulation Language）两种类型的初始化。DDL是对表结构的创建、修改或删除，DML是建表后对表内数据的增删改。DML初始化其实很常见，比如，对系统中的角色表或管理员信息的初始化，或者一些字典性的业务数据。

Spring Boot默认会从类路径的根目录下加载名为schema.sql和data.sql的两个脚本文件，这两个脚本文件分别对应DDL脚本和DML脚本。

初始化脚本可以通过spring.sql.init.schema-locations和spring.sql.init.data-locations两个配置项来分别指定。这两个配置项支持指定多个初始化脚本，当指定多个时，默认使用分隔符";"来分隔。这里的分隔符还可以通过配置项spring.sql.init.separator来自定义。比如可以配置如下：

```
spring:
  sql:
    init:
      separator: ","
      schema-locations: "classpath:ddl1.sql,classpath:ddl2.sql"
      data-locations: "classpath:dml1.sql,classpath:dml2.sql"
```

Spring Boot还支持根据不同的数据库来指定不同的初始化脚本，使用配置项spring.sql.init.platform来指定数据库的类别。比如配置如下：

```
spring:
  sql:
    init:
      platform: "mysql"
```

如此一来，默认情况下Spring Boot读取的DDL和DML脚本名就成了schema-mysql.sql和data-mysql.sql。

对于使用脚本初始化，如果数据有安全设置，比如项目中数据库连接的角色不具有DDL权限等情况，此时可以通过spring.sql.init.username和spring.sql.init.password配置项，单独设置初始化脚本所使用的数据库用户名和密码。

除了上面提及的配置项外，Spring Boot还支持两个初始化脚本相关的配置项，下面分别说明。

- spring.sql.init.continue-on-error：用来指定当初始化脚本执行中遇到错误时是否继续执行。此配置项默认值为false，即初始化脚本执行错误后停止执行脚本。如果需要在遇到错误时继续执行，可以将此配置项设置为true。
- spring.sql.init.encoding：指定DDL和DML脚本的编码。

9.4.2 使用JPA和Hibernate时初始化数据库

在前面的章节中，我们已经使用过Spring Data JPA在启动时自动初始化数据库的功能。Spring Data JPA默认使用Hibernate JPA作为实现，所以Spring Data JPA和Hibernate在使用配置上也有相似之处。

Spring Data JPA和Hibernate JPA可以根据实体类中关于表结构的定义自动生成DDL语句，应用在程序启动时执行，以初始化数据库。

使用Spring Data JPA时，可以通过配置项spring.jpa.generate-ddl来开启或关闭应用程序启动时初始化数据库的功能。此配置项默认是false，也就是默认初始化数据库的功能是关闭的。

另外，还有一个专门用于Hibernate JPA作为JPA实现时的配置项spring.jpa.hibernate.ddl-auto，用来设置DDL是否执行以及执行策略。此配置项有5个配置值可选，分别是none、validate、update、create和create-drop。

配置项spring.jpa.hibernate.ddl-auto值的选取一定要小心，避免误删数据。在这5个配置值中，none不会执行任何操作；validate只会检测当前实体类中定义的表结构和数据库中的表结构是否一致，如果不一致，则会报错；update也是常用的值，这个值会比较数据库中的表结构和实体类中的表结构是否一致，如果不一致，则会创建表、修改表、创建索引等；create和create-drop都会在启用时创建表，这两个值的不同之处是create会在创建表之前先删除当前存在的表，而create-drop是在关闭sessionFactory时自动删除创建的表。create和create-drop两个属性值都有删除表的动作，通常只会在测试或者开发环境中使用。

Spring Boot会根据当前的检测结果来为配置项spring.jpa.hibernate.ddl-auto设置属性值。如果没有检测到模式管理器，则配置值为create-drop，在所有其他情况下配置值都为none。

此外，如果配置项spring.jpa.hibernate.ddl-auto被设置为create或create-drop，则在启动时执行类路径根目录下名为import.sql的文件。

第 10 章

项目实战1——客户管理Web系统

前面的章节介绍了MyBatis、Spring Security、Actuator、日志框架等多种技术和Spring Boot的整合，本章将通过开发客户管理Web系统来展示在一个项目中对多种技术的整合使用。

本章主要涉及的知识点有：

- ⌘ 了解客户管理Web系统的需求。
- ⌘ 了解开发所需要的相关技术。
- ⌘ 如何使用Spring Initializr初始化项目。
- ⌘ 如何实现客户管理Web系统。

10.1 梳理业务需求

要开发一个系统，首先需要梳理清楚需求。本节将说明客户管理Web系统具体的需求。

顾名思义，客户管理Web系统是一个基于Web的应用程序，用来对客户的信息进行管理。对于客户信息管理，首先需要有客户信息，所以应该有客户信息的录入功能。而录入信息的具体实现方式（有表单录入、爬虫爬取、Excel导入等），在本例系统中只实现表单录入的方式。

客户有哪些信息，有哪些信息是需要录入的，这些信息需要梳理清楚，具体列举在表10.1中。

表 10.1 客户的信息

信　　息	信息来源	说　　明
ID	系统生成	在录入时系统生成
客户名称	管理员录入	
类别	管理员录入	字典值：个人、企业、非营利机构
联系电话	管理员录入	
电子邮箱	管理员录入	

(续表)

信　　息	信息来源	说　　明
客户地址	管理员录入	
所属区域	管理员录入	字典值：华北、东北、华东、华南、西南、西北和港澳台地区
客户经理	管理员录入	
登记日期	管理员录入	
最后一次交易时间	系统生成	最后一次发货的时间
备注	管理员录入	备注信息

对于客户的信息录入后，还需要能查看到当前系统中录入的用户信息。展示信息的形式通常是列表，在此系统中需要有一个页面用来展示客户列表。客户列表中展示表10.1中列举出的全部字段信息。

在客户列表页面还需要有搜索框，用来根据客户名称进行搜索，以及支持根据客户类别和所属区域进行筛选。还有一个"编辑"按钮，用来打开编辑页面，在编辑页面中可以编辑除ID、最后一次交易时间以外的所有信息，并保存。

客户关联交易的信息，比如销售额、销售时间等也是构成客户信息的一部分，最后一次的发货信息作为客户的"最后一次交易"。

客户关联的交易信息需要有一个独立的交易信息录入页，包括交易信息展示列表、交易信息的编辑以及标注为已发货。具体交易信息列举在表10.2中。

表 10.2　交易信息

信　　息	信息来源	说　　明
ID	系统生成	在录入时系统生成
商品名称	管理员录入	
总数量	管理员录入	
交易总价	管理员录入	
下单时间	管理员录入	
是否发货	管理员录入	字典项，包括未发货、已发货
发货时间	管理员录入	

10.2　技术实现设计

分析前面提到的需求，首先可以定位项目为B/S架构，涉及数据的存储以及数据的增加、查询和修改操作。

为了满足上述需求，可以选择Spring MVC作为Web框架，选择MySQL作为数据库。对于数据访问层，选择众多，Spring Data JPA、MyBatis和Hibernate等都可以，这里我们选择前面介绍过但是没有实战的MyBatis。

由于这是一个管理系统，因此还要考虑访问安全问题。在前面的实践项目中使用过Shiro作为认证授权框架，这里我们不妨使用Spring Security作为认证授权框架。

关于数据库，我们可以根据需求的描述设计出系统的表结构。表的设计可以有多种，这里的设计是客户信息和交易信息分别对应一张表。另外，客户信息和交易信息是一对多的关系，因此不妨将映射关系保存在交易信息表中，即交易信息表中保存客户ID。

客户信息表设计如表10.3所示。

表 10.3　客户信息表

字　　段	数据类型	说　　明
id	INT	使用自增序列生成
cname	VARCHAR(50)	客户名称
category	VARCHAR(15)	客户类别。字典值：个人、企业、非营利机构
phone_number	VARCHAR(20)	客户联系电话
email	VARCHAR(100)	客户电子邮箱
address	VARCHAR(500)	客户地址
region	VARCHAR(25)	客户所属区域。字典值：华北、东北、华东、华南、西南、西北和港澳台地区
manager	VARCHAR(50)	客户经理
registration_date	DATE	客户信息的登记日期
last_deal_time	DATETIME	最后一次发货的时间
remark	VARCHAR(500)	备注信息

交易信息表设计如表10.4所示。

表 10.4　交易信息表

字　　段	字段类型	说　　明
id	INT	使用自增序列生成
gname	VARCHAR(50)	商品名称
count	INT	总数量
total	FLOAT	交易总价
deal_time	DATETIME	下单时间
has_sent	CHAR(6)	是否发货
send_time	DATETIME	发货时间
cid	INT	客户ID

项目后端的设计大致为以上内容，前端的实现并不作为重点，前端页面的具体实现只是采用HTML和jQuery技术，不涉及复杂的样式。

10.3　构　建　项　目

10.3.1　使用 Spring Initializr 构建项目

在梳理清楚项目需求和完成项目设计之后，开始项目的开发工作。

（1）首先构建项目，构建项目的方式有很多，在这里使用Spring Initializr作为构造器。在浏览器中访问https://start.spring.io/，如图10.1所示。

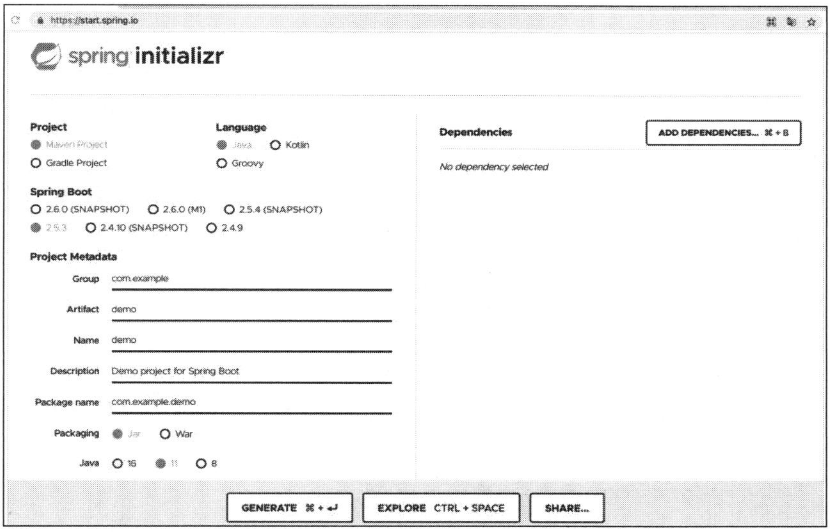

图 10.1 Spring Initializr 初始页面

使用默认的Maven作为项目构建工具，语言默认使用Java，Spring Boot版本使用当前的GA 2.5.3版本。项目信息中的Artifact修改为customermanager，Name和包名（Package Name）使用自动生成的包名。简单修改Description，对项目进行简要说明。

对于其他依赖信息，项目使用MySQL作为数据库，所以需要MySQL驱动。使用MyBatis作为数据访问层，所以需要MyBatis依赖，以及使用Spring MVC作为Web框架，Spring Security作为安全框架，所以添加对应的依赖，最后启用Spring Boot Actuator作为项目管理的工具。具体配置如图10.2所示，最后单击GENERATE按钮，生成项目的压缩包并下载。

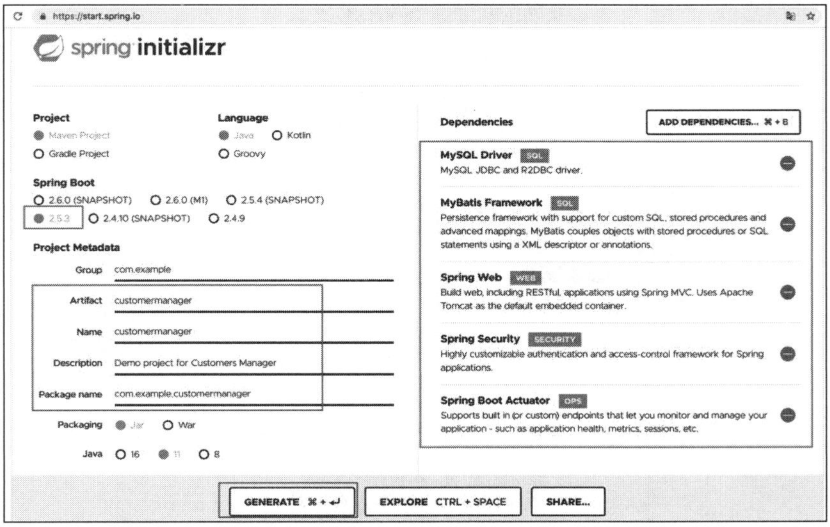

图 10.2 在 Spring Initializr 中配置项目

默认生成的压缩包是以Artifact命名、ZIP类型的，如图10.3所示。

图 10.3 下载的项目压缩包

（2）将其解压并将解压的文件移动到工作空间目录，比如这里以目录IdeaCProjects为工作空间，解压后的customermanager与工作空间如图10.4所示。

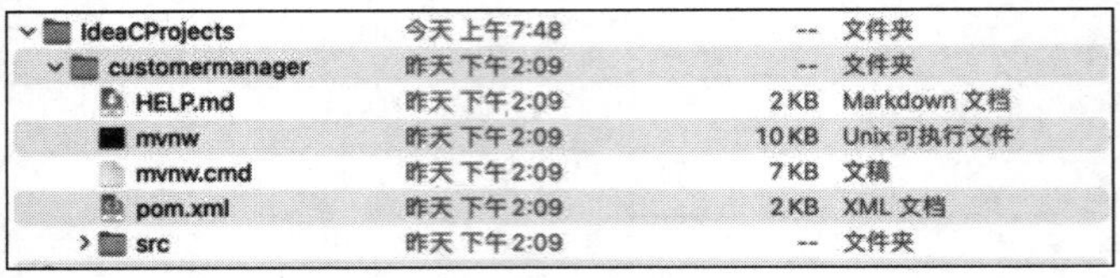

图 10.4 工作空间和项目的目录结构

（3）接下来使用IDE加载customermanager项目，这里选用IDEA作为IDE。

打开IDEA，如图10.5所示。

图 10.5 IDEA 初始页面

依次选择Projects→Open，打开目录选择窗口，如图10.6所示。

图 10.6 选择自己的项目

选择项目目录，并单击Open按钮，打开IDEA的工作窗口，如图10.7所示。

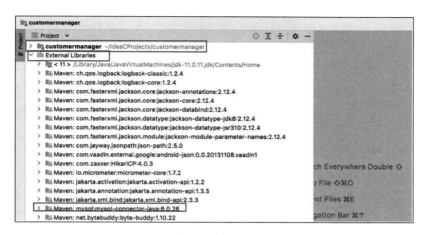

图 10.7　项目工作目录结构

（4）在窗口中可以看到当前加载的项目目录，正是工作空间IdeaCProjects下的项目customermanager。当Maven对项目解析完成之后，可以在External Libraries部分看到对项目所添加的依赖。打开customermanager部分，如图10.8所示，可以看到项目目录下的所有文件。

双击pom.xml，在右侧编辑器窗口中打开pom.xml文件，如图10.9所示。从文件中可以看到我们在Spring Initializr中所添加的依赖。

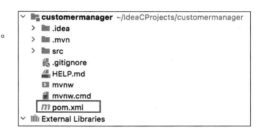

图 10.8　项目的文件列表

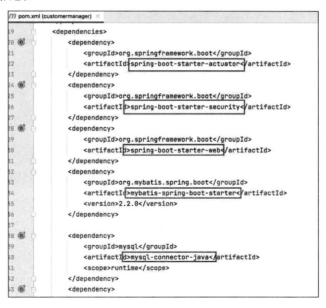

图 10.9　项目依赖

在项目开发中，主要通过编辑文件内容来对依赖进行管理。

（5）项目的构建工作到这里就基本完成了。运行项目可以通过在文件CustomermanagerApplication.java上右击，选中"Run 'Customermanage....main()'"来启用，如图10.10所示。

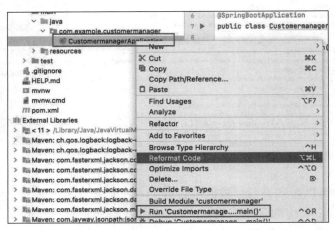

图 10.10 启动项目

不过此时启动项目会报错，错误信息如图10.11所示。

错误原因是当前项目中既没有内置数据库，又没有指定外部数据库的信息。下一节继续处理这个问题。

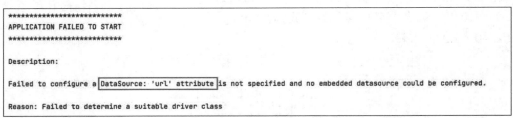

图 10.11 未配置数据库资源时报错

10.3.2 配置数据库

接下来指定数据库的地址、用户名和密码。项目中在resources目录下有一个默认生成的application.properties配置文件，文件内容为空。我们使用YAML格式的配置文件，所以将application.properties文件重命名为application.yaml。然后双击打开application.yaml，并在其中配置数据库的连接信息，具体配置代码如下：

```
spring:
  datasource:
    url: "jdbc:mysql://127.0.0.1:3306/demo"
    username: "root"
    password: "123456"
```

上面是作者本机的数据库地址、用户名和密码，具体配置应以用户自己的数据库连接信息为准。

配置完成后，再次启动项目，控制台输出如图10.12所示，项目启动成功。

```
vletWebServerApplicationContext : Root WebApplicationContext: initialization completed in 2586 ms
DetailsServiceAutoConfiguration :

e3c07

.web.EndpointLinksResolver      : Exposing 1 endpoint(s) beneath base path '/actuator'
.DefaultSecurityFilterChain     : Will secure any request with [org.springframework.security.web.context.reque
embedded.tomcat.TomcatWebServer : Tomcat started on port(s): 8080 (http) with context path ''
tomermanagerApplication         : Started CustomermanagerApplication in 5.139 seconds (JVM running for 5.794)
```

图 10.12　项目启动成功的日志输出

10.4　创建数据库表

根据10.2节的分析，可以设计出两张数据库表，分别是客户信息表和交易信息表。根据字段信息定义出建表语句，客户信息表建表语句如下：

```
CREATE TABLE customers (
  id int NOT NULL AUTO_INCREMENT,
  cname varchar(50) CHARACTER SET utf8 COLLATE utf8_general_ci NOT NULL COMMENT '客户名称。',
  category varchar(15) CHARACTER SET utf8 COLLATE utf8_general_ci DEFAULT NULL COMMENT '客户类别。字典值：个人、企业、非营利机构。',
  phone_number varchar(20) CHARACTER SET utf8 COLLATE utf8_general_ci DEFAULT NULL COMMENT '客户联系电话。',
  email varchar(100) CHARACTER SET utf8 COLLATE utf8_general_ci DEFAULT NULL COMMENT '客户电子邮箱。',
  address varchar(500) CHARACTER SET utf8 COLLATE utf8_general_ci DEFAULT NULL COMMENT '客户地址。',
  region varchar(25) CHARACTER SET utf8 COLLATE utf8_general_ci DEFAULT NULL COMMENT '客户所属区域。字典值：华北、东北、华东、华南、西南、西北和港澳台地区。',
  manager varchar(50) CHARACTER SET utf8 COLLATE utf8_general_ci DEFAULT NULL COMMENT '客户经理。',
  registration_date date DEFAULT NULL COMMENT '客户信息的登记日期。',
  last_deal_time varchar(45) DEFAULT NULL COMMENT '最后一次发货的时间。',
  remark varchar(500) CHARACTER SET utf8 COLLATE utf8_general_ci DEFAULT NULL COMMENT '备注信息。',
  PRIMARY KEY (id),
  UNIQUE KEY cname_UNIQUE (cname)
) ENGINE=InnoDB DEFAULT CHARSET=utf8mb3;
```

交易信息表建表语句如下：

```
CREATE TABLE deals (
  id int NOT NULL AUTO_INCREMENT,
  gname varchar(50) CHARACTER SET utf8 COLLATE utf8_general_ci DEFAULT NULL COMMENT '商品名称。',
  count int DEFAULT NULL COMMENT '总数量。',
  total float DEFAULT NULL COMMENT '交易总价。',
```

```
    deal_time datetime DEFAULT NULL COMMENT '下单时间。',
    has_sent char(6) DEFAULT NULL COMMENT '是否发货。',
    send_time datetime DEFAULT NULL COMMENT '发货时间。',
    cid int DEFAULT NULL COMMENT '客户ID。',
    PRIMARY KEY (id)
) ENGINE=InnoDB DEFAULT CHARSET=utf8mb3;
```

通过数据库工具连接数据库，执行以上建表语句，创建出客户信息表customers和交易信息表deals。

10.5 开发客户信息模块

根据需求和设计，客户信息模块需要有列表页面、添加页面、修改页面。对于页面的入口，在系统首页上放置"客户列表"和"添加客户"按钮，分别用来打开客户列表页面和添加客户页面。在客户列表中放置"修改"按钮，用来打开"修改客户信息"页面。接下来逐个开发页面和对应的接口。

10.5.1 开发系统首页

在项目目录resources/static下创建index.html作为系统首页，设置title为"客户信息管理Web系统-主页"，并创建列表，指定两个超链接"添加客户"和"客户列表"作为列表项。超链接的地址先空着，具体index.html的代码如下：

```
<!DOCTYPE html>
<html lang="en">
<head>
    <meta charset="UTF-8">
    <title>客户信息管理 Web 示例系统-主页</title>
</head>
<body>
<h1>客户信息管理系统</h1>
<ul>
    <li>
        <a href="">添加客户</a>
    </li>
    <li>
        <a href="">客户列表</a>
    </li>
</ul>
</body>
</html>
```

由于当前系统中已经启用了Spring Security，为了访问方便，选择配置用户和密码来代替默认的用户和随机密码，配置用户名为admin，密码为123456，在application.yaml中添加配置代码如下：

```yaml
spring:
  security:
    user:
      name: admin
      password: 123456
```

重启/启动项目，在浏览器中访问地址http://127.0.0.1:8080/，自动跳转到Spring Security的登录页面，如图10.13所示。

录入我们前面设置的用户名admin和密码123456，单击Sign in按钮。如果是错误页面则不会跳转，并且提示错误信息。登录成功后，页面会自动跳转到项目首页，也就是我们前面所写的index.html，浏览器显示如图10.14所示。

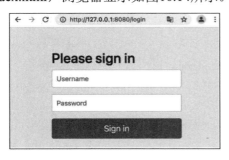

图10.13　默认登录页面

图10.14　项目首页

10.5.2　开发添加客户页面和接口

页面应该包含9个文本框，分别对应除ID和最后一次交易时间之外的其他9个字段。

1. 开发添加客户信息接口

创建类CustomerController，客户信息相关的接口都放置到这个类中。首先定义添加客户信息接口，设计思路是在接口中保存数据到数据库，并返回生成的ID，而保存数据需要使用MyBatis的功能。

（1）创建Customer实体类，用来和数据库表customers进行映射。Customer类参考表字段设计，部分代码如下：

```java
package com.example.customermanager.pojo;

import java.sql.Date;
import java.sql.Timestamp;
import java.util.Objects;

public class Customer {
    private Integer id;
    private String cname;
    private String category;
    private String phoneNumber;
    private String email;
    private String address;
    private String region;
```

```
    private String manager;
    private Date registrationDate;
    private Timestamp lastDealTime;
    private String remark;
    public Customer() {
    }
    // 这里省略了有参数构造, get、set 方法和 equeals、hash 方法等
}
```

(2) 创建DAO接口和Mapper映射文件。

创建dao目录,并创建CustomerMapper接口,使用@Repository注解和@Mapper注解标注,并在其中定义添加方法,具体代码如下:

```
package com.example.customermanager.dao;

import com.example.customermanager.pojo.Customer;
import org.apache.ibatis.annotations.Mapper;
import org.springframework.stereotype.Repository;

@Repository
@Mapper
public interface CustomerMapper {
    int add(Customer customer);
}
```

在resources目录下创建mapper目录,并在目录下创建CustomerMapper.xml mapper文件,文件代码如下:

```
<?xml version="1.0" encoding="UTF-8"?>
<!DOCTYPE mapper PUBLIC "-//mybatis.org//DTD Mapper 3.0//EN"
"http://mybatis.org/dtd/mybatis-3-mapper.dtd">
<mapper namespace="com.example.customermanager.dao.CustomerMapper">
    <insert id="add" parameterType="com.example.customermanager.pojo.Customer">
        insert into customers(cname, category, phone_number,
            email, address, region,
            manager, registration_date, remark)
        values (#{cname}, #{category}, #{phoneNumber},
            #{email}, #{address}, #{region},
            #{manager}, #{registrationDate}, #{remark})
    </insert>
</mapper>
```

然后在application.yaml配置文件中配置mapper文件的扫描路径,具体配置代码如下:

```
mybatis:
  mapper-locations: "classpath:mapper/*.xml"
```

(3) 创建controller目录,在目录下创建CustomerController,并在类中定义接口,已经自动注入CustomerMapper接口的实例。CustomerController的具体代码如下:

```
package com.example.customermanager.controller;
import com.example.customermanager.dao.CustomerMapper;
import com.example.customermanager.pojo.Customer;
import org.springframework.beans.factory.annotation.Autowired;
import org.springframework.web.bind.annotation.PutMapping;
import org.springframework.web.bind.annotation.RequestMapping;
import org.springframework.web.bind.annotation.RestController;
@RestController
@RequestMapping("customer")
public class CustomerController {
    @Autowired
    CustomerMapper mapper;
    @PutMapping("add")
    public String add(Customer customer){
        return "" + mapper.add(customer);
    }
}
```

通过上面的代码可以看出，新增客户信息的接口地址为customer/add，方法类型为put，参数为Customer实体类类型，这些信息在开发页面时需要用到。

2. 开发添加客户信息页面

在客户信息页面中，需要录入除客户ID和最后交易时间之外的9个字段和一个保存按钮。这9个字段中，客户类别、客户所属区域使用下拉选择框，登记日期使用日期选择框，其余6个都使用文本输入框。在单击提交时，通过jQuery获取用户录入的信息，并通过Ajax调用接口。

具体来讲，先在static目录下创建customer目录，后面所有只关于客户信息的页面都放置到customer目录下。在customer目录下创建add_customer.html文件，并实现具体的逻辑，部分代码如下：

```
<!DOCTYPE html>
<body>
<h1>客户信息管理系统-添加客户</h1>
<div>
    <table>
        <tr>
            <td>客户名称</td>
            <td><input id="cname" type="text"></td>
        </tr>
        <tr>
            <td>客户类别</td>
            <td>
                <select id="category">
                    <option>个人</option>
                    <option>企业</option>
                    <option>非盈利机构</option>
                </select>
            </td>
```

```html
            </tr>
    ……
        </table>
    </div>
    <script>
    function save(){
        var remark = $("#remark").val();
        var registration_date = $("#registration_date").val();
        var manager = $("#manager").val();
        var region = $("#region").val();
        var address = $("#address").val();
        var email = $("#email").val();
        var phone_number = $("#phone_number").val();
        var category = $("#category").val();
        var cname = $("#cname").val();
        var param_data = {"remark":remark, "registrationDate":registration_date,
"manager":manager,
            "region":region, "address":address, "email":email,
            "phoneNumber":phone_number, "category":category, "cname":cname };
        $.ajax({
            type: 'PUT',
            url: '/customer/add',
            withCredentials: true,
            data: param_data,
            success: function(data){
                alert("保存客户信息成功！");
            },
            error: function(data){
                // 失败后执行的代码
                alert("保存客户信息失败！");
                console.log(status);
                return;
            }
        });
    }
    </script>
    </body>
    </html>
```

3. 配置 Spring Security 停用跨域限制

Spring Security默认开启了跨域校验，会拦截PUT、POST等请求，所以为了正常访问接口，需要禁用跨域限制。创建config目录，在config目录下创建SecurityConfig配置类，并继承WebSecurityConfigurerAdapter类，重写config方法，具体代码如下：

```
package com.example.customermanager.config;

import org.springframework.context.annotation.Configuration;
import org.springframework.security.config.annotation.web.builders.HttpSecurity;
import org.springframework.security.config.annotation.web.configuration.
```

WebSecurityConfigurerAdapter;

```
@Configuration
public class SecurityConfig extends WebSecurityConfigurerAdapter {
    @Override
    protected void configure(HttpSecurity http) throws Exception {
        super.configure(http);
        http.csrf().disable();
    }
}
```

4. 在首页中配置添加客户信息页面的路径

前面开发了系统首页，当时添加客户信息的页面路径被空出来了，现在修改主页面代码。添加客户信息页面相对于首页的路径为customer/add_customer.html，所以修改index.html文件第11行代码，修改后代码如下：

```
<a href="customer/add_customer.html">添加客户</a>
```

此时，项目中的文件如图10.15所示。

其中，框起来的是在本节所创建的文件或目录。

5. 访问页面和接口

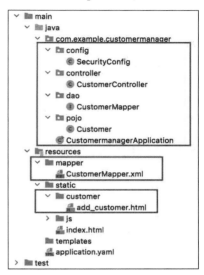

图10.15　新增加的文件和目录

启动项目，在浏览器访问地址http://127.0.0.1:8080/，自动跳转到登录页面，输入用户名和密码登录。跳转到系统首页，此时系统首页中的"添加客户"链接已经指定了添加客户页面的地址。单击"添加客户"链接，自动跳转到添加客户页面，如图10.16所示。

录入列表中各字段的信息，如图10.17所示，并单击"保存"按钮。

图10.16　添加客户页面　　　　　　　　　图10.17　添加客户录入信息

页面提示保存客户信息成功，如图10.18所示。

登录MySQL数据库，查询数据库中的数据，如图10.19所示，和前面在页面中录入的信息一致，说明页面和接口运行正常。

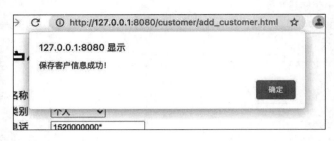

图 10.18　保存客户信息成功提示窗口

图 10.19　数据库查询增加的客户信息

10.5.3　开发客户列表页面和接口

列表页面需要在列表中渲染数据，这里引入Thymeleaf进行数据渲染。首先在项目中引入Thymeleaf依赖，即在pom.xml中添加如下代码：

```
<dependency>
  <groupId>org.springframework.boot</groupId>
  <artifactId>spring-boot-starter-thymeleaf</artifactId>
</dependency>
```

1．开发客户列表接口

在引入Thymeleaf之后，需要对原先的CustomerController做一些改造。由于需要在CustomerController类中增加列表接口，列表接口会返回template名称，因此之前的@RestController需要修改成@Controller，并且为了保证之前的添加客户信息接口正常返回字符串结果，需要使用@ResponseBody注解标注接口方法add(Customer customer)。改造完成之后，开始开发列表接口。

接口因为有查询、页码、分页大小，所以应该有3个参数。同时考虑到查询结果的封装，所以增加一个Map类型的参数，用于封装返回值。封装参数时，因为有查询条件和分页，所以在封装查询结果时还需要封装分页信息。

对于查询部分，仍然会使用到MyBatis。在接口CustomerMapper中定义查询方法，具体方法定义如下：

```
List<Customer> find(String keyword, int offset, int size);
```

然后在CustomerMapper.xml文件中创建find方法对应的查询语句，具体的查询语句代码如下：

```
<select id="find" resultType="com.example.customermanager.pojo.Customer">
    select id, cname, category, phone_number as phoneNumber,
    email, address, region,
    manager, registration_date as registrationDate, last_deal_time as lastDealTime, remark
    from customers
```

```xml
<where>
<if test="keyword != null">
    cname like #{keyword}
</if>
</where>
limit #{offset}, #{size}
</select>
```

再回到CustomerController中完善接口方法，接口方法代码如下：

```java
@RequestMapping
public String listPage(@RequestParam(required = false, defaultValue = "") String keyword,
                @RequestParam(required = false, defaultValue = "0") int page,
                @RequestParam(required = false, defaultValue = "10") int size,
                Map<String, Object> map) {
    String keywordQuery = ("".equals(keyword.trim()) ? null : ("%" + keyword.trim() + "%"));
    List<Customer> customers = customerMapper.find(keywordQuery, page * size, size);
    map.put("customers", customers);
    map.put("keyword", keyword);
    map.put("pageLast", page == 0 ? 0 : page - 1);
    map.put("pageNext", page + 1);
    map.put("size", size);
    return "customer_list";
}
```

2. 开发客户列表页面

列表页面分成两部分，一部分是搜索框和查询按钮，另一部分是列表。列表又分为表头、数据部分和分页按钮（链接）。列表需要显示需求部分列举的11个字段，在第12列提供一个编辑的链接，用来打开编辑当前行数据的编辑页面。页面的开发需要使用到Thymeleaf的属性头。

页面使用Thymeleaf模板来开发，于是这里使用默认的目录template，在template目录下创建customer_list.html文件，并在文件中完成客户列表页面的功能，页面的具体代码如下：

```html
<!DOCTYPE html>
<body>
<h2>客户信息列表</h2>
<hr>
<input id="keyword" type="text" th:value="${keyword}"/>
<button id="search">查询</button>
<hr>
<table style="border:1px solid #ccc" rules="rows">
  <thead>
  <tr>
    <th>客户 ID</th>
    <th>客户名称</th>
    <th>客户类别</th>
    <th>客户电话</th>
    <th>客户邮箱</th>
```

```html
        <th>客户地址</th>
        <th>所属区域</th>
        <th>客户经理</th>
        <th>登记时间</th>
        <th>最后一次交易时间</th>
        <th>备注</th>
        <th>
        </th>
      </tr>
    </thead>
    <tbody id="list">
    <tr th:each="customer:${customers}">
      <td th:text="${customer.id}"></td>
      <td th:text="${customer.cname}"></td>
      <td th:text="${customer.category}"></td>
      <td th:text="${customer.phoneNumber}"></td>
      <td th:text="${customer.email}"></td>
      <td th:text="${customer.address}"></td>
      <td th:text="${customer.region}"></td>
      <td th:text="${customer.manager}"></td>
      <td th:text="${customer.registrationDate}"></td>
      <td th:text="${customer.lastDealTime}"></td>
      <td th:text="${customer.remark}"></td>
      <td>
        <a th:href="@{'/customer/customer_edit?id='+${customer.id}}">编辑</a>
      </td>
    </tr>
    </tbody>
    <tfoot>
      <tr>
        <td>
          <a th:href="@{'/customer?keyword=' + ${keyword} + '&page=' + ${pageLast} + '&size=' + ${size}}">上一页</a>
        </td>
        <td>
          <a th:href="@{'/customer?keyword=' + ${keyword} + '&page=' + ${pageNext} + '&size=' + ${size}}">下一页</a>
        </td>
      </tr>
    </tfoot>
  </table>
  <script>
  $('#search').click(function(){
    window.location.href = "/customer?keyword="+$("#keyword").val();
  })
  </script>
</body>
</html>
```

3. 在浏览器中访问客户列表页面和接口

为了开发中测试方便，修改SecurityConfig的配置代码，设置所有的请求都允许未登录访问，配置代码如下：

```
@Configuration
public class SecurityConfig extends WebSecurityConfigurerAdapter {
    @Override
    protected void configure(HttpSecurity http) throws Exception {
        http.csrf().disable();
        http.authorizeRequests().anyRequest().permitAll();
    }
}
```

启动项目，在浏览器中访问系统首页http://127.0.0.1:8080/，没有登录直接进入系统首页。然后单击客户列表链接，浏览器跳转到客户列表页面，如图10.20所示。

图 10.20　客户信息列表页面

可以看到，页面展示包括在上一节中添加的那条数据。不妨回到系统首页，选择添加客户，跳转到添加用户页面，录入信息，如图10.21所示，单击"保存"按钮。

图 10.21　添加客户页面录入信息

提示保存信息成功后回退到系统主页，然后再次单击"客户列表"链接，浏览器跳转到客户列表页面，此时展示出了两条信息，如图10.22所示。

图10.22　添加客户后的客户列表页面

可以在查询输入框中输入"实业",然后单击"查询"按钮,页面刷新。注意浏览器地址栏的变化,可以观察到此时只查询到客户名称带有"实业"的第二条数据(id为5),如图10.23所示。

图10.23　客户列表查询结果

单击"下一页"链接,页面刷新,查询不到数据,如图10.24所示,因为查询keyword为实业的数据只有一条。

至此,完成了客户信息列表页面和接口的开发。列表页面中虽然有"编辑"超链接,但是"编辑"页面和接口尚未完成,这是接下来要开发的功能。

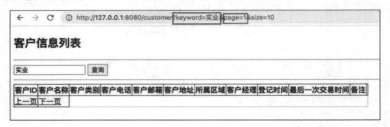

图10.24　客户信息列表页第二页

10.5.4　开发编辑客户信息页面和接口

因为页面中需要加载数据,所以编辑页面和接口依然考虑使用Thymeleaf模板引擎。

1. 开发编辑客户信息接口

这里其实需要两个接口,一个是打开编辑页面的接口,另一个是保存修改后数据信息的接口。

打开编辑页面的接口地址已经在客户列表页面customer_list.html中指定了,为"customer/customer_edit?id="。这个接口中的逻辑就是根据参数id查询客户信息,并将信息返回指定的

模板页面。创建接口方法openEditPage并实现查询的逻辑,接口方法的具体代码如下:

```
@GetMapping("customer_edit")
public String openEditPage(@RequestParam int id, Map<String, Object> map){
    Customer customer = customerMapper.findById(id);
    map.put("customer", customer);
    return "customer_edit";
}
```

上面的方法返回之后,框架会去找名为customer_edit.html的模板文件解析。

下面开发客户数据更新接口。首先开发数据访问层,在CustomerMapper接口中定义update方法,定义代码如下:

```
int update(Customer customer);
```

然后在CustomerMapper.xml文件中编写更新的SQL语句,注意last_deal_time字段是系统修改的,不支持用户修改,所以这里更新时不更新last_deal_time字段。具体的更新SQL语句代码如下:

```
<update id="update" parameterType="com.example.customermanager.pojo.Customer">
    update customers
    set cname=#{cname}, category=#{category}, phone_number=#{phoneNumber},
    email=#{email}, address=#{address}, region=#{region},
    manager=#{manager}, registration_date=#{registrationDate},
    remark=#{remark}
    where id=#{id}
</update>
```

再回到CustomerController类中编写客户信息更新接口。接口使用Customer实体类接口参数,然后调用CustomerMapper接口的update方法保存,最后使用@ResponseBody标注表示返回的不是页面,具体代码如下:

```
@PostMapping("update")
@ResponseBody
public String update(Customer customer) {
    if(customer.getId()==null){
        return "0";
    }
    return "" + customerMapper.update(customer);
}
```

至此,完成了接口的开发。

2. 开发编辑客户信息页面

根据接口的定义,页面需要在template目录下创建名为customer_edit.html的文件。页面在加载时应该显示除last_deal_time外的10个字段,并且设置id文本框禁止修改。在单击"保存"按钮后,使用jQuery获取文本框中的数据,然后通过Ajax调用update接口。具体代码如下:

```
<!DOCTYPE html>
<body>
```

```html
<h1>客户信息管理系统-编辑客户信息</h1>
<div>
  <table>
    <tr>
      <td>客户ID</td>
      <td><input id="id" type="test" disabled th:value="${customer.id}"></td>
    </tr>
    <tr>
      <td>客户名称</td>
      <td><input id="cname" type="text" th:value="${customer.cname}"></td>
    </tr>
    <tr>
      <td>客户类别</td>
      <td>
        <select id="category" th:value="${customer.category}">
          <option>个人</option>
          <option>企业</option>
          <option>非盈利机构</option>
        </select>
      </td>
    </tr>
    ...
  </table>
</div>
<script>
function save(){
    var remark = $("#remark").val();
    var registration_date = $("#registration_date").val();
    var manager = $("#manager").val();
    var region = $("#region").val();
    var address = $("#address").val();
    var email = $("#email").val();
    var phone_number = $("#phone_number").val();
    var category = $("#category").val();
    var cname = $("#cname").val();
    var id = $("#id").val();
    var param_data = {"remark":remark,
        "registrationDate":registration_date, "manager":manager,
        "region":region, "address":address, "email":email,
        "phoneNumber":phone_number, "category":category, "cname":cname, "id":id};
    $.ajax({
        type: 'POST',
        url: '/customer/update',
        withCredentials: true,
        data: param_data,
        success: function(data){
            alert("更新客户信息成功!");
        },
        error: function(data){
```

```
                // 失败后执行的代码
                alert("更新客户信息失败！");
                console.log(status);
                return;
            }
        });
    }
</script>
</body>
</html>
```

3. 测试编辑客户信息

在浏览器中访问http://127.0.0.1:8080/进入系统首页，然后单击"客户列表"链接进入客户列表页面。此时列表中的"编辑"链接已经可以打开编辑页面，比如单击数据ID为1的"编辑"链接，页面跳转到编辑客户信息窗口，并且在页面中显示当前的客户信息，如图10.25所示。

编辑字段的信息，如图10.26所示，并单击"更新"按钮。

图 10.25　编辑客户信息页面　　　　　　　　图 10.26　修改客户信息

页面提示更新成功，如图10.27所示。

图 10.27　更新客户信息成功提示窗口

回到数据库中，查询数据，数据库中的数据如图10.28所示。

图 10.28　数据查询修改后的客户信息

id为1的数据更新为页面新录入的数据，并且另一条数据没有受影响，所以更新数据成功。单击"确定"按钮后，使用浏览器的回退按钮回到客户列表页，可以看到更新后的数据，如图10.29所示。

图 10.29　查看客户信息列表页

10.6　开发交易信息模块

交易信息模块与客户信息模块相比，除了共有的新增、列表、修改之外，还增加了标注为已发货的功能，目的是在标注为已发货时更新客户信息中最后一次交易的时间字段。

10.6.1　在系统首页增加交易信息导航

在系统首页的链接列表中增加"创建交易"和"查询交易"两个链接，分别用于打开创建交易页面和交易列表页面。

仿照客户信息的两个链接，在index.html中添加如下代码：

```
<li>
    <a href="deal/add_deal.html">创建交易</a>
</li>
<li>
    <a href="deal">交易列表</a>
</li>
```

启动项目，在浏览器中访问首页http://127.0.0.1:8080/，已经有了"创建交易"和"交易列表"两个链接，如图10.30所示。

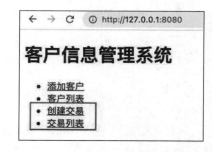

图 10.30　系统首页新增两个链接

10.6.2　开发"创建交易"页面和接口

"创建交易"页面类似于"添加客户"页面，不用考虑数据的加载，所以没有必要使用模板引擎，使用静态页面即可。

1. 定义交易实体类

根据技术实现设计中的数据类型，定义交易实体类。在pojo目录下创建Deal类，并创建字段、有参无参构造、getter、setter等，具体代码如下：

```
package com.example.customermanager.pojo;

import java.sql.Timestamp;
import java.util.Objects;

public class Deal {
    private Integer id;
    private String gname;
    private Integer count;
    private Float total;
    private Timestamp dealTime;
    private String hasSent;
    private Timestamp sendTime;
    private Integer cid;

    public Deal() {
    }
// 这里省略了有参构造、toString、getter、setter 等方法
}
```

2. 创建交易数据访问层

在dao目录下创建DealMapper接口，并定义其中的方法。由于已经有了开发客户信息模块的基础，因此DealMapper可以复制接口CustomerMapper，并将其中的Customer（或customer）全部修改为Deal（或deal）。此时，DealMapper接口的代码如下：

```
package com.example.customermanager.dao;

import com.example.customermanager.pojo.Deal;
import org.apache.ibatis.annotations.Mapper;
import org.springframework.stereotype.Repository;

import java.util.List;

@Repository
@Mapper
public interface DealMapper {
    int add(Deal deal);

    List<Deal> find(String keyword, int offset, int size);
    Deal findById(int id);
    int update(Deal deal);
}
```

接下来创建MyBatis的mapper文件，也可以将CustomerMapper.xml复制一份，并将副本重命名为DealMapper.xml。对DealMapper.xml代码进行清洗，将其中的表名customers全部修改为交易记录的表名deals，并修改实体类以及具体字段的内容。经过代码清洗之后，DealMapper.xml的代码如下：

```xml
<?xml version="1.0" encoding="UTF-8"?>
<!DOCTYPE mapper PUBLIC "-//mybatis.org//DTD Mapper 3.0//EN"
"http://mybatis.org/dtd/mybatis-3-mapper.dtd">
<mapper namespace="com.example.customermanager.dao.DealMapper">
    <insert id="add" parameterType="com.example.customermanager.pojo.Deal">
        insert into deals(gname, count, total, deal_time, has_sent, cid)
        values (#{gname}, #{count},
            #{total}, #{dealTime}, #{hasSent},
            #{cid})
    </insert>
    <select id="find" resultType="com.example.customermanager.pojo.Deal">
        select id, gname, count, total,
            deal_time as dealTime, has_sent as hasSent,
            send_time as sendTime, cid
        from deals
        <where>
        <if test="keyword != null">
            gname like #{keyword}
        </if>
        </where>
        limit #{offset}, #{size}
    </select>
    <select id="findById" resultType="com.example.customermanager.pojo.Deal">
        select id, gname, count, total,
            deal_time as dealTime, has_sent as hasSent,
            send_time as sendTime, cid
        from deals
        where id=#{id}
    </select>
    <update id="update" parameterType="com.example.customermanager.pojo.Deal">
        update deals
        set gname=#{gname}, count=#{count}, total=#{total},
            deal_time=#{dealTime}
        where id=#{id}
    </update>
</mapper>
```

进行数据插入时不维护send_time字段，等到标注为已发货时再维护send_time字段。

至此，完成了数据访问层的创建。

3. 开发创建交易接口

在创建交易接口时，仍然可以借助已经开发好的客户信息模块的接口来实现。复制类CustomerController，并将副本重命名为DealController。然后对DealController中的代码进行清洗，将Customer相关的类和接口修改为Deal相关的类和接口，这样DealController中就定义好了增加、查询和修改相关的接口，具体代码如下：

```java
package com.example.customermanager.controller;

import com.example.customermanager.dao.DealMapper;
import com.example.customermanager.pojo.Deal;
import org.springframework.beans.factory.annotation.Autowired;
import org.springframework.stereotype.Controller;
import org.springframework.web.bind.annotation.*;

import java.util.List;
import java.util.Map;

@Controller
@RequestMapping("Deal")
public class DealController {
    @Autowired
    DealMapper dealMapper;

    @PutMapping("add")
    @ResponseBody
    public String add(Deal deal) {
        return "" + dealMapper.add(deal);
    }

    @RequestMapping
    public String listPage(@RequestParam(required = false, defaultValue = "") String keyword,
                           @RequestParam(required = false, defaultValue = "0") int page,
                           @RequestParam(required = false, defaultValue = "10") int size,
                           Map<String, Object> map) {
        String keywordQuery = ("".equals(keyword.trim()) ? null : ("%" + keyword.trim() + "%"));
        List<Deal> deals = dealMapper.find(keywordQuery, page * size, size);
        map.put("deals", deals);
        map.put("keyword", keyword);
        map.put("pageLast", page == 0 ? 0 : page - 1);
        map.put("pageNext", page + 1);
        map.put("size", size);
        return "deal_list";
    }

    @GetMapping("deal_edit")
    public String openEditPage(@RequestParam int id, Map<String, Object> map){
        Deal deal = dealMapper.findById(id);
        map.put("deal", deal);
        return "deal_edit";
    }

    @PostMapping("update")
    @ResponseBody
    public String update(Deal deal) {
        if(deal.getId()==null){
            return "0";
        }
```

```
        return "" + dealMapper.update(deal);
    }
}
```

4. 开发创建交易页面

在static下创建deal目录，然后复制add_customer.html到新建的deal目录下，并将副本修改为add_deal.html。接下来需要对add_deal.html文件中的代码进行清洗，根据设置的字段定义出文本框，并在单击"保存"按钮时获取数据。对于下单时间需要特殊处理，在用户录入数据之后拼接":00.0"以匹配后台的数据类型。代码经过清洗后，add_deal.html的代码如下：

```
<!DOCTYPE html>
<body>
<h1>客户信息管理系统-创建交易</h1>
<div>
    <table>
        <tr>
            <td>商品名称</td>
            <td><input id="gname" type="text"></td>
        </tr>
...
        <tr>
            <td>是否发货</td>
            <td>
                <select id="hasSent" disabled>
                    <option>未发货</option>
                    <option>已发货</option>
                </select>
            </td>
        </tr>
        <tr>
            <td>所属的客户ID</td>
            <td><input id="cid" type="text"></td>
        </tr>
        <tr>
            <td colspan="2">
                <button onclick="save(this)">保存</button>
            </td>
        </tr>
    </table>
</div>
<script>
function save(){
    var gname = $("#gname").val();
    var count = $("#count").val();
    var total = $("#total").val();
    var dealTime = $("#dealTime").val();
    dealTime = (dealTime ? dealTime.replace("T", " ") +":00.0" : dealTime)
    var hasSent = $("#hasSent").val();
    var cid = $("#cid").val();
```

```
            var param_data = {"gname":gname , "count":count , "total":total ,
                "dealTime":dealTime , "hasSent":hasSent , "cid":cid };
            $.ajax({
                type: 'PUT',
                url: '/deal/add',
                withCredentials: true,
                data: param_data,
                success: function(data){
                    alert("创建交易成功！");
                },
                error: function(data){
                    // 失败后执行的代码
                    alert("创建交易失败！");
                    console.log(status);
                    return;
                }
            });
        }
    </script>
    </body>
</html>
```

注意　在页面中是否发货字段默认为未发货，并禁止在此页面修改。

5. 测试创建交易页面和接口

在浏览器中访问系统首页，地址为http://127.0.0.1:8080/。单击"创建交易"链接，浏览器跳转到创建交易页面，如图10.31所示。

在页面中录入字段信息，如图10.32所示。

图 10.31　创建交易页面　　　　　　　图 10.32　录入交易信息

单击"保存"按钮，提示创建交易成功，如图10.33所示。

图 10.33 创建交易信息成功提示窗口

10.6.3 开发"交易列表"页面

在前面清洗DealController和DealMapper代码时,已经完成了交易列表的接口部分,所以这里只需要开发页面。仍然选择基于客户信息列表页进行修改,在template目录下复制customer_list.html,并将副本重命名为deal_list.html。接下来对deal_list.html中的代码进行清洗。

我们需要将一些静态信息修改为交易的信息,对于列表,需要修改列表的字段为交易的字段。代码经过清洗后,deal_list.html的部分代码如下:

```html
<!DOCTYPE html>
<body>
<h2>交易列表</h2>
<hr>
<input id="keyword" type="text" th:value="${keyword}"/>
<button id="search">查询</button>
<hr>
<table style="border:1px solid #ccc" rules="rows">
  <thead>
  <tr>
    <th>交易ID</th>
    <th>商品名称</th>
    <th>总数量</th>
    <th>总金额</th>
    <th>交易时间</th>

……
</table>
<script>
$('#search').click(function(){
  window.location.href = "/deal?keyword="+$("#keyword").val();
})
</script>
</body>
</html>
```

启动项目,在浏览器中访问系统首页http://127.0.0.1:8080/,然后单击"交易列表"链接,跳转到交易列表页,页面如图10.34所示。

图 10.34　交易列表页面

10.6.4　开发"编辑交易"页面

同样，这里只需要开发页面。复制template目录下的customer_edit.html文件，并将副本重命名为deal_edit.html，然后对代码进行清洗。清理代码的思路和前面大致相似，只是需要对时间类型数据的展示进行特殊处理。经过清洗后，deal_edit.html模板页面的代码如下：

```
<!DOCTYPE html>
<body>
<h1>客户信息管理系统-编辑交易信息</h1>
<div>
  <table>
    <tr>
      <td>交易ID</td>
      <td><input id="id" type="text" disabled th:value="${deal.id}"></td>
    </tr>
    <tr>
      <td>商品名称</td>
      <td><input id="gname" type="text" th:value="${deal.gname}"></td>
    </tr>
    <tr>
      <td>总数量</td>
      <td><input id="count" type="text" th:value="${deal.count}"></td>
    </tr>
    <tr>
      <td>总金额</td>
      <td><input id="total" type="text" th:value="${deal.total}"></td>
    </tr>
    <tr>
      <td>交易时间</td>
      <td><input id="dealTime" type="datetime-local" th:value=
"${#dates.format(deal.dealTime,'yyyy-MM-dd')}+'T'+${#dates.format(deal.dealTime,'HH:mm:ss')}"></td>
    </tr>
    ……
  </table>
</div>
<script>
function save(){
    var id = $("#id").val();
```

```
        var gname = $("#gname").val();
        var count = $("#count").val();
        var total = $("#total").val();
        var dealTime = $("#dealTime").val();
        dealTime = (dealTime ? dealTime.replace("T", " ") +(dealTime.length==16? ":00.0" :".0") : dealTime)
        var param_data = {"id":id, "gname":gname , "count":count , "total":total ,
            "dealTime":dealTime };
        $.ajax({
            type: 'POST',
            url: '/deal/update',
            withCredentials: true,
            data: param_data,
            success: function(data){
                alert("更新交易信息成功！");
            },
            error: function(data){
                // 失败后执行的代码
                alert("更新交易信息失败！");
                console.log(status);
                return;
            }
        });
    }
</script>
</body>
</html>
```

启动项目，在浏览器中访问系统主页，然后进入"交易列表"页，再单击"编辑"链接，跳转到编辑交易信息页面，页面展示如图10.35所示。

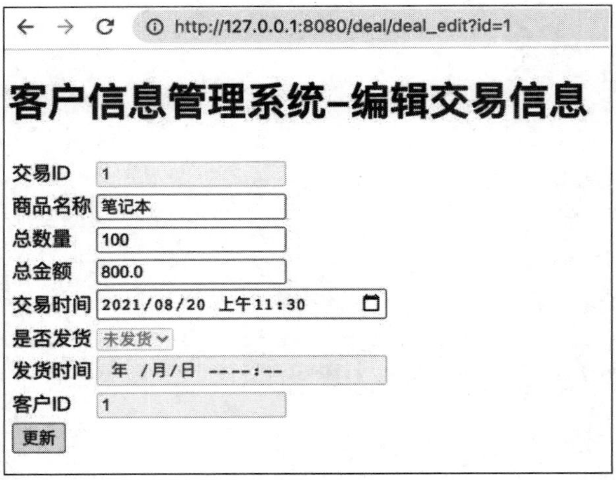

图 10.35　编辑交易信息页面

将总金额800.0修改成900.0，然后单击"更新"按钮。页面提示更新成功，从数据库查看，如图10.36所示，可以看到total已经修改成了900。

id	gname	count	total	deal_time	has_sent	send_time	cid
1	笔记本	100	900	2021-08-20 11:30:00	未发货	NULL	1

图 10.36　数据库查看修改后的交易信息

10.6.5　开发标注发货状态功能

该功能的效果是在交易列表最后一栏，当未发货时有一个按钮"已发货"，单击该按钮后，按钮所在行的交易记录被更新成已发货，并且更新关联客户的最后一次交易时间，更新成功后刷新当前页面。

1. 完成数据访问层的接口开发

在标注已发货的接口中需要交易信息表和客户信息表，需要做两次更新。

先定义更新交易信息的接口。这里根据交易id更新，更新has_sent字段为"已发货"，更新send_time为当前时间，所以接口参数为id。在DealMapper接口中定义方法，具体代码如下：

```
int send(int id);
```

然后在mapper文件DealMapper.xml中完善方法对应的SQL语句，添加update标签，具体代码如下：

```xml
<update id="send" >
   update deals
   set has_sent='已发货',
       send_time=now()
   where id=#{id}
</update>
```

再来定义更新客户信息的接口方法，这里是根据客户信息id来更新的，方法定义的代码如下：

```
int updateLastDealTime(int id);
```

最后在mapper文件CustomerMapper.xml中完善SQL语句，先创建update标签，具体代码如下：

```xml
<update id="updateLastDealTime">
   update customers
   set last_deal_time=now()
   where id=#{id}
</update>
```

2. 开发标注已发货功能接口

因为这是基于交易的操作，所以仍然在DealController类中添加方法。方法中需要更新客户信息的字段，所以在类中注入CustomerMapper。定义接口方法，在接口方法中获取客户id，通过交易表做一次查询得到。接口方法不是返回页面，所以需要@ResponseBody注解标注，具体代码如下：

```
@Autowired
CustomerMapper customerMapper;
@PostMapping("send")
@ResponseBody
public String send(Integer id){
    dealMapper.send(id);
    Deal deal = dealMapper.findById(id);
    Integer cid = deal.getCid();
    customerMapper.updateLastDealTime(cid);
    return "1";
}
```

3. 修改交易列表页面，增加标注发货的按钮

在列表体内的最后加上一列，并用th:if标注，条件为当hasSent为"未发货"时显示此列，然后内部放置一个按钮。HTML代码如下：

```
<td th:if="${deal.hasSent} eq '未发货'">
    <button class="sendBtn" th:did="${deal.id}">标注为发货</button>
</td>
```

使用jQuery给按钮绑定单击事件，当单击按钮时，访问标注发货接口，并在接口响应成功后刷新当前页面，具体JS代码如下：

```
$('.sendBtn').click(function(){
    let did=$(this).attr("did");
    $.ajax({
        type: 'POST',
        url: '/deal/send',
        withCredentials: true,
        data: {"id":did},
        success: function(data){
            alert("标注发货成功");
            window.reload();
        },
        error: function(data){
            // 失败后执行的代码
            alert("标注发货失败！");
            console.log(status);
            return;
        }
    });
})
```

4. 测试标注发货的按钮

启动项目，访问系统首页http://127.0.0.1:8080/。为了测试新创建一条交易信息，表单填写如图10.37所示。

单击"保存"按钮，提示"创建交易成功"。然后回到系统首页，再进入交易列表页，此时交易列表页如图10.38所示。

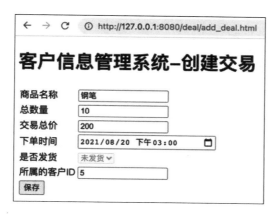

图 10.37　创建交易

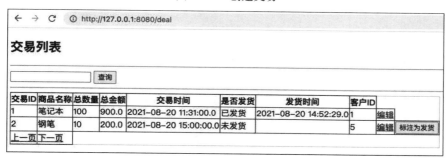

图 10.38　交易列表页

第二条数据为刚添加的记录，状态为未发货，显示标注按钮。单击标注按钮，页面提示标注成功，如图10.39所示。

图 10.39　标注成功提示窗口

单击"确定"按钮，页面自动刷新，第二条数据的是否发货变更成了"已发货"，并且去掉了"标注为发货"按钮，如图10.40所示。

图 10.40　交易列表

最后回退到系统主页，再进入客户信息列表页面，此时id为5的客户信息中最后一次交易时间已经更新，如图10.41所示。

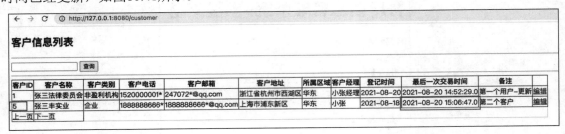

图10.41 查看客户最后一次交易时间

第 11 章

项目实战2——个人博客

在上一章实现了客户管理系统，本章来实现一个可以用来公开访问的个人博客系统，支持用户的注册、登录，以及评论、点赞等功能。希望通过本章的实战使读者巩固前面学到的Spring Boot基础知识。

本章主要涉及的知识点有：

- ⌘ 了解个人博客示例系统的需求。
- ⌘ 了解开发个人博客示例系统所需要的具体技术。
- ⌘ 如何使用Spring Initializr初始化项目。
- ⌘ 项目的模块划分与代码实现。

11.1 梳理业务需求

首先来梳理清楚业务需求，说明个人博客示例系统要做哪些事。本节将梳理个人博客示例系统具体的需求。

个人博客系统，目的是做成一个可以发布到互联网上，允许互联网用户注册并访问的博客系统。相比于如今网络上成熟的博客平台，功能简单许多。

个人博客示例系统具有发布博客的功能，不过限于篇幅，仅支持管理员（系统所有者）发布博客，因此博客列表也就是管理员个人的博客列表，这样实现起来相对简单。当博客被浏览时，系统记录博客的浏览次数加1。还支持用户单击喜欢和取消喜欢一篇博客的功能。梳理出博客主体具体包含的信息列举在表11.1中。

表 11.1 博客主体包含的信息

信　　息	信息来源	说　　明
ID	系统生成	创建博客时每一篇博客应该有一个唯一id
标题	管理员录入	博客的标题

(续表)

信　　息	信息来源	说　　明
发布时间	管理员录入	发布博客的日期、时间，格式如：2021年08月30日19:30
博客内容	管理员录入	博客的文本信息
浏览次数	程序维护	每加载一次博客此字段数值加1
喜欢次数	程序维护	用户可以单击喜欢和取消喜欢

系统允许互联网用户公开访问，用户可以注册、登录、修改密码等。系统预置管理员账号和密码，用于发布博客。互联网用户使用手机号+验证码登录，首次登录就是注册。登录以后可以通过验证码来修改密码，设置密码后用户可以通过密码登录。梳理出用户主体包含的信息如表11.2所示。

表 11.2　用户主体包含的信息

信　　息	信息来源	说　　明
ID	系统生成	用户注册时每人应该有一个唯一id
手机号	用户录入	用户使用手机号+验证码登录，首次登录默认注册。手机号在系统中应该唯一
密码	用户录入	修改密码时生成。用户可以通过手机号验证码修改密码
注册时间	系统生成	用户注册时设置此字段
更新时间	系统生成	用户修改信息时更新此字段

对于登录的用户，可以在个人首页修改密码等。

11.2　技术实现设计

分析前面提到的需求，首先可以定位项目为B/S架构，涉及数据的存储以及数据的增加、查询、删除和修改操作。

为了满足上述需求，可以选择Spring MVC作为Web框架，数据库选择MySQL作为数据库。对于数据访问层，这里选择使用Spring Data JPA来实现。为实现用户登录和权限控制，项目中引入Spring Security框架来实现。页面技术采用静态页面和Thymeleaf动态模板基础结合的方式来实现。

下面分模块设计页面、接口和数据库表结构。

11.2.1　博客模块

根据需求的数据可以设计出数据库博客表的字段信息，具体信息列举在表11.3中。

表 11.3　博客表的设计

字　段　名	字段类型	说　　明
id	INT	自增序列
title	VARCHAR(200)	博客标题

（续表）

字 段 名	字段类型	说　　明
publish_time	DATETIME	发布时间
content	TEXT	博客内容
scan_count	INT	浏览次数
like_count	INT	喜欢次数

项目初始化内置管理员账号，管理员可以发布博客，所以应该有发布博客的页面和发布博客的接口。发布博客的页面包含表11.4中列举的字段信息。

表 11.4　发布博客页面的字段信息

信　　息	实现形式	说　　明
博客标题	Input	用于用户录入博客标题信息
博客内容	textarea	用于录入博客内容

发布博客页面底部有"发布"按钮，单击该按钮后调用"发布博客"接口。接口定义在本节最后说明。

博客列表页面也是系统的主页面，以分页的方式展示博客列表。列表每条记录显示博客的标题、创建时间、浏览次数和喜欢次数。页面使用Thymeleaf模板引擎技术实现，接口和页面地址统一，接口具体信息在本节最后说明。

此外还有博客详情页面。博客详情页面显示博客标题、博客创建时间、博客浏览次数、喜欢次数和博客内容，在博客内容下面是评论列表。评论列表也支持分页。因为博客详情页面包含分页的评论列表，所以使用Thymeleaf模板引擎技术并不是很方便，于是采用HTML和Ajax的方式来实现。

在博客详情页面还有必要增加对一个接口的调用，就是增加博客的浏览次数。另外，还有由用户在图书详情页来触发的接口，就是"喜欢/取消喜欢博客"接口，这个接口在当前用户没有喜欢时为"喜欢"，在已经喜欢时为"取消喜欢"。本节涉及的所有接口列举在表11.5中。

表 11.5　博客模块接口梳理

接　口　名	接口地址	Method	参　　数	返　回　值
发布博客	/blog	PUT	title：博客标题。content：博客内容	1表示发布成功，0表示发布失败
博客列表	/blog	GET	page：页码，默认从0开始 size：每页大小	博客列表页面
博客详情	/blog/{id}	GET	id：博客的id	博客实体类
浏览次数计数	/blog/{id}/scan	POST	id：博客的id	返回增加后的浏览次数

11.2.2　用户模块

根据需求可以定义出用户主体数据库表的字段信息，列举在表11.6中。

表 11.6 用户表的设计

字 段 名	字段类型	说　　明
id	INT	自增序列
phone_number	VARCHAR(20)	用户手机号
password	VARCHAR(100)	存储加密后的密码
regi_time	DATETIME	用户注册时间
update_time	DATETIME	用户信息最后一次修改时间

用户登录页面有验证码登录和密码登录。验证码登录需要有"发送验证码"接口，密码登录需要有"密码登录"接口。密码登录需要先设置密码，有"通过验证码设置密码"接口。

验证码登录需要有手机号输入文本框、验证码输入文本框以及"发送验证码"按钮。单击"发送验证码"按钮调用"发送验证码"接口。单击登录按钮调用"验证码登录"接口。

修改密码页面有当前用户手机号、验证码和带设置的密码3个文本输入框，以及"发送验证码"按钮和"提交"按钮。单击"发送验证码"按钮调用"验证码登录"接口。单击"提交"按钮调用"修改密码"接口。

密码登录有手机号文本框和密码文本框，单击"登录"按钮调用"密码登录"接口。

本节涉及的接口列举在表11.7中。

表 11.7　用户模块接口梳理

接　口　名	接口地址	Method	参数	返　回　值
发送验证码	/login/sms	PUT	phone_number：手机号	1：成功，0：失败
验证码登录	/login/byvc	POST	phone_number：手机号 verify_code：验证码	token：用来设置到请求头中，作为登录标识
修改密码	/user/chpw	POST	verify_code：验证码 password：密码	1：成功；0：失败
密码登录	/login/bypw	POST	phone_number：手机号 password：密码	token：用来设置到请求头中，作为登录标识

11.2.3　喜欢、取消喜欢博客功能

为了实现"喜欢博客"和"取消喜欢博客"接口，应该记录哪些用户喜欢博客，所以还需要一张表记录用户喜欢博客的数据。用户喜欢博客记录表的表字段设计列举在表11.8中。

表 11.8　用户喜欢博客表字段设计

字 段 名	字段类型	说　　明
id	INT	自增序列
blog_id	INT	博客id
user_id	INT	用户id
like_time	DATETIME	喜欢时间

模块接口设计列举在表11.9中。

表11.9 博客模块接口梳理

接 口 名	接口地址	Method	参 数	返 回 值
喜欢博客	/blog/{id}/like	POST	id：博客的id	成功返回1，不成功返回0
取消喜欢博客	/blog/{id}/dislike	POST	id：博客的id	成功返回1，不成功返回0

11.3 构 建 项 目

根据前面技术实现的梳理，需要在项目中使用Spring MVC、Thymeleaf、Spring Security、Spring Data JPA和MySQL，由此可以梳理出项目需要添加的依赖。此外，为了免去写getter、setter、构造器、toString和hashCode等方法，引入Lombok。

仍然使用Spring Initializr来初始化项目，在浏览器中访问https://start.spring.io/，并在页面中配置项目元信息、项目依赖信息等，配置如图11.1所示。

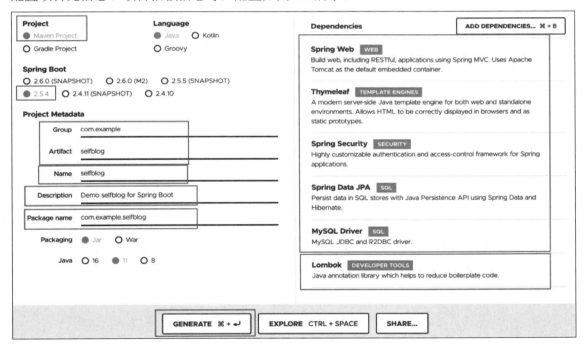

图11.1 Spring Initializr 构建项目

配置完成后单击GENERATE按钮，将生成selfblog.zip文件并保存到本地。

对于IDE没有限制，下面以IDEA Community为例进行介绍。将压缩包解压到IDEA项目目录，如图11.2所示。

然后打开IDEA，选择File→NEW→Project from Existing Sources...，如图11.3所示。

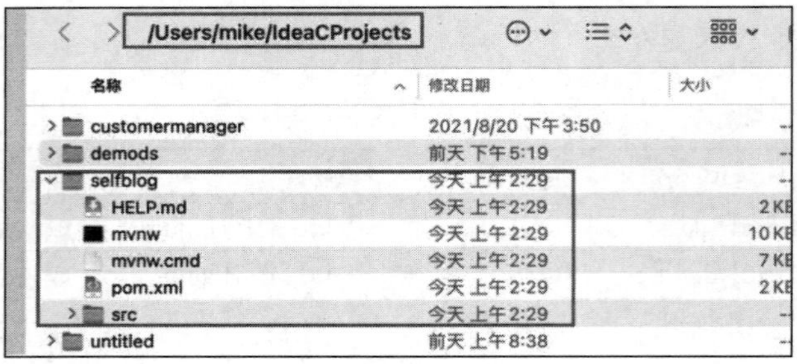

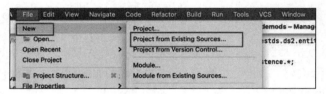

图 11.2　项目解压后的目录

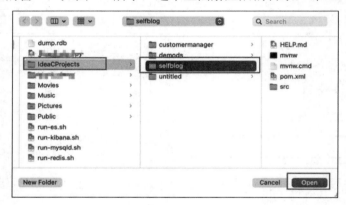

图 11.3　使用 IDEA 打开项目

打开选择目录的窗口，如图11.4所示，选中上面解压后的目录，即selfblog目录。

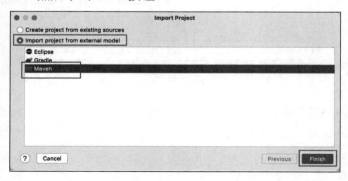

图 11.4　选择解压后的项目目录

单击Open按钮，打开Import Project窗口，如图11.5所示，选择Import project from external model，再选择Maven，然后单击Finish按钮。

图 11.5　选择导入项目的构建工具

在Open Project窗口单击This Window按钮，如图11.6所示。

图 11.6　在当前窗口打开项目

因为项目引入了Spring Data JPA，在项目启动时会初始化数据库，不配置数据库连接信息项目无法启动，所以先在配置文件中配置数据库连接，配置文件在resources目录下，默认是application.properties，重命名为application.yaml。配置代码如下：

```
spring:
  datasource:
    url: "jdbc:mysql://127.0.0.1:3306/selfblog"
    username: "root"
    password: "123456"
```

启动项目，可以在看见控制台启动成功的日志输出，如图11.7所示。

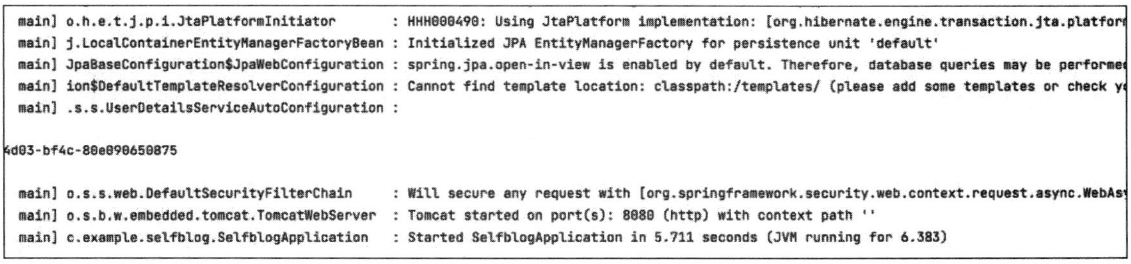

图 11.7　启动项目在控制台的日志输出

Spring Security配置了默认的登录页面，在浏览器中访问地址http://127.0.0.1:8080/login，打开登录页面，如图11.8所示，说明项目构建完成，可以正常运行。

图 11.8　Spring Security 默认登录页面

11.4　创建数据实体类

在技术设计中已经梳理出项目需要的数据库表，可以编写SQL建表语句来创建表，但是这种编写SQL语句的方式不够灵活，在表结构有变化时，既需要修改实体类又需要修改SQL语句，尤其是在有多套环境时，操作烦琐且容易出错。

Spring Data JPA可以根据实体类定义自动生成建表语句，启动项目时可以自动初始化数据库，并且在实体类定义修改时，可以自动对数据库表结构进行修改。所以在本节创建实体中使用Spring Data JPA初始化数据库。

根据博客表的设计可以定义出实体类Blog，代码如下：

```java
package com.example.selfblog.entity;

import lombok.AllArgsConstructor;
import lombok.Data;
import lombok.NoArgsConstructor;

import javax.persistence.Column;
import javax.persistence.Entity;
import javax.persistence.Id;
import javax.persistence.Table;
import java.sql.Timestamp;

@Data
@AllArgsConstructor
@NoArgsConstructor
@Entity
@Table(name = "blogs")
public class Blog {
    @Id
    @GeneratedValue(strategy = GenerationType.IDENTITY)
    private Integer id;

    @Column(columnDefinition = "varchar(200)")
    private String title;

    private Timestamp publishTime;

    @Column(columnDefinition = "text")
    private String content;

    private Integer scanCount;

    private Integer likeCount;
}
```

@Entity、@Table、@Id和@Column都是JPA中定义的注解，表名通过注解@Table的name属性指定为"blogs"。

使用@Data、@AllArgsConstructor和@NoArgsConstructor注解，以提供类的get、set、equals、hashCode、canEqual、toString方法及全量参数构造器和无参数构造器。

接下来，根据用户表的定义设计实体类User，代码如下：

```java
package com.example.selfblog.entity;

import lombok.AllArgsConstructor;
import lombok.Data;
import lombok.NoArgsConstructor;
```

```java
import javax.persistence.Column;
import javax.persistence.Entity;
import javax.persistence.Id;
import javax.persistence.Table;
import java.sql.Timestamp;

@Data
@AllArgsConstructor
@NoArgsConstructor
@Entity
@Table(name = "users")
public class User {
    @Id
    @GeneratedValue(strategy = GenerationType.IDENTITY)
    private Integer id;

    @Column(columnDefinition = "varchar(20)")
    private String phoneNumber;

    @Column(columnDefinition = "varchar(100)")
    private String password;

    private Timestamp regiTime;

    private Timestamp updateTime;
}
```

用户表除了字段与博客表不同之外，涉及的技术基本一样。

最后还有一个跨模块的实体类，用户喜欢博客实体类，代码如下：

```java
package com.example.selfblog.entity;

import lombok.AllArgsConstructor;
import lombok.Data;
import lombok.NoArgsConstructor;

import javax.persistence.*;
import java.sql.Timestamp;

@Data
@AllArgsConstructor
@NoArgsConstructor
@Entity
@Table(name = "user_liked_blog")
public class UserLikedBlog {
    @Id
    @GeneratedValue(strategy = GenerationType.IDENTITY)
    private Integer id;

    @ManyToOne(optional = true)
    private Blog blog;

    @ManyToOne(optional = true)
    private User user;
```

```
    private Timestamp likeTime;
}
```

在配置文件中指定DDL初始化的策略，配置代码如下：

```
spring:
  jpa:
    hibernate:
      ddl-auto: "update"
      show-sql: "true"
```

配置完成之后启动项目，日志输出可以看到3条建表语句，如图11.9所示。

```
Hibernate: create table blogs (id integer not null auto_increment, content text, like_count integer,
    publish_time datetime(6), scan_count integer, title varchar(200), primary key (id)) engine=InnoDB
Hibernate: create table user_liked_blog (id integer not null auto_increment, like_time datetime(6),
    blog_id integer, user_id integer, primary key (id)) engine=InnoDB
Hibernate: create table users (id integer not null auto_increment, password varchar(100), phone_number
    varchar(20), regi_time datetime(6), update_time datetime(6), primary key (id)) engine=InnoDB
```

图11.9 建表语句日志输出

由于在实体类中使用了@ManyToOne注解，这些注解用来关联表的字段，在生成DDL语句时会被处理为对应的外键约束，因此在日志输出中，紧接着建表语句之后可以看到添加外键约束的两条DDL语句，如图11.10所示。

```
Hibernate: alter table user_liked_blog add constraint FKkspbq5o7n61ft86kl8o11j3ru foreign key (blog_id)
    references blogs (id)
Hibernate: alter table user_liked_blog add constraint FK5a7ftv6csafe9pd99rypymdqm foreign key (user_id)
    references users (id)
```

图11.10 创建索引的语句输出

在数据库中，可以查看到新创建的3张表blogs、user_liked_blog和users，如图11.11所示。

图11.11 在数据库可视化工具中查看创建的表

至此，便完成了实体类以及数据库表的创建。

11.5 开发博客模块

开发博客模块也是从接口入手，再开发相关的页面和数据访问层。

为了调试方便，先将Spring Security设置为允许所有请求，并且接口能正常访问，禁用跨域限制，具体配置代码如下：

```
@Configuration
public class Config extends WebSecurityConfigurerAdapter {
    @Override
    protected void configure(HttpSecurity http) throws Exception {
        http.authorizeRequests().anyRequest().permitAll();
        http.csrf().disable(); // 禁用csrf
    }
}
```

接下来逐个开发设计的接口。首先创建controller包，在controller包下创建BlogController类。因为使用了Thymeleaf模板引擎，部分接口需要返回页面，所以controller类上使用@Controller注解，而对于不是返回页面的接口方法，使用@ResponseBody注解标注，并在类中注入BlogRepository以便于后续接口使用。BlogController类的代码如下：

```
package com.example.selfblog.controller;

import org.springframework.stereotype.Controller;
import org.springframework.web.bind.annotation.PutMapping;
import org.springframework.web.bind.annotation.RequestMapping;

import java.sql.Timestamp;

@Controller
@RequestMapping("blog")
public class BlogController {
    @Autowired
    BlogRepository blogRepository;
}
```

BlogRepository接口的代码如下：

```
package com.example.selfblog.repository;

import com.example.selfblog.entity.Blog;
import org.springframework.data.jpa.repository.JpaRepository;
import org.springframework.data.jpa.repository.JpaSpecificationExecutor;

public interface BlogRepository extends JpaRepository<Blog, Integer>, JpaSpecificationExecutor<Blog> {
}
```

11.5.1 开发发布博客接口和页面

发布博客页面不需要加载数据，所以使用HTML+jQuery就可以完成页面，并使用Ajax调用发布博客接口完成发布功能。

先来开发接口。根据接口定义，在BlogController类中定义publishBlog方法，使用注解@PutMapping标注为PUT类型请求，使用实体类Blog作为参数，如果成功就返回1，具体代码如下：

```
@PutMapping
@ResponseBody
public String publishBlog(@RequestBody Blog blog){
```

```
            blog.setPublishTime(new Timestamp(System.currentTimeMillis()));
            blog.setLikeCount(0);
            blog.setScanCount(0);
            blogRepository.save(blog);
            return "1";
    }
```

再来开发页面。页面结构简单,页面中只需要有"标题"文本框、"内容"文本框和"保存"按钮。"标题"文本框使用input组件,"内容"文本框使用textarea组件,并为按钮绑定事件,在单击"保存"按钮时调用发布博客接口,页面的具体代码如下:

```html
<!DOCTYPE html>
<body>
<h1>个人博客系统-发布博客</h1>
<div>
    <table>
        <tr>
            <td>博客标题</td>
            <td><input id="title" type="text"></td>
        </tr>
        <tr>
            <td>博客内容</td>
            <td><textarea id="content" rows="20" cols="60"></textarea></td>
        </tr>
        <tr>
            <td colspan="2">
                <button onclick="publish(this)">保存</button>
            </td>
        </tr>
    </table>
</div>
<script>
function publish(){
    var title = $("#title").val();
    var content = $("#content").val();
    var param_data = {"title" : title, "content" : content };
    $.ajax({
        type: 'PUT',
        url: '/blog',
        contentType: 'application/json',
        withCredentials: true,
        data: JSON.stringify(param_data),
        success: function(data){
            alert("发布博客成功!");
        },
        error: function(data){
            // 失败后执行的代码
            alert("发布博客失败!");
            console.log(data);
            return;
```

```
        }
    });
}
</script>
</body>
</html>
```

至此，便完成了发布博客页面和接口的开发，现在可以使用页面来发布博客。启动项目，在浏览器中访问地址http://127.0.0.1:8080/publish_blog.html，页面如图11.12所示。

在页面中录入标题和博客内容，如图11.13所示。

图 11.12　发布博客页面

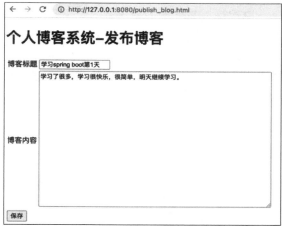

图 11.13　录入博客信息

然后单击"保存"按钮，页面提示发布博客成功，如图11.14所示。

图 11.14　发布博客成功提示窗口

从数据库中查询blogs表的数据，可以发现新生成的一条博客记录，如图11.15所示。

图 11.15　在数据库查看生成的博客记录

存在从页面录入的title、content和生成的发布时间等数据，说明发布博客功能开发完成。为了方便博客列表的开发和测试，再发布5条博客，创建的数据如图11.16所示。

id	content	like_count	publish_time	scan_count	title
1	学习了很多，学习很快乐，很简…	0	2021-08-24 20:07:40.660000	0	学习spring boot第1天
2	学习第二天：学习了很多，学习…	0	2021-08-24 20:12:03.312000	0	学习spring boot第2天
3	学习第三天：学习了很多，学习…	0	2021-08-24 20:12:10.566000	0	学习spring boot第3天
4	学习第四天：学习了很多，学习…	0	2021-08-24 20:12:17.236000	0	学习spring boot第4天
5	学习第五天：学习了很多，学习…	0	2021-08-24 20:12:23.690000	0	学习spring boot第5天
6	学习第六天：学习了很多，学习…	0	2021-08-24 20:12:31.272000	0	学习spring boot第6天

图 11.16　博客表的全部记录

11.5.2　开发博客列表接口和页面

博客列表需要加载博客数据，采用Thymeleaf模板来实现。通过接口查询数据，然后数据返回模板渲染页面并返回给浏览器。

在BlogController中用注解@GetMapping定义blogs方法，这是一个接收GET请求的接口。定义map以及分页参数page和size，查询调用blogRepository的findAll方法。接口的具体代码如下：

```java
@GetMapping
public String blogs(@RequestParam(required = false, defaultValue = "0") int page,
                    @RequestParam(required = false, defaultValue = "5") int size,
                    Map<String, Object> map){
    Page<Blog> blogs = blogRepository.findAll(PageRequest.of(page, size));
    map.put("blogs", blogs.getContent());
    map.put("pageLast", page == 0 ? 0 : page - 1);
    map.put("pageNext", page + 1);
    map.put("size", size);
    map.put("total", blogs.getTotalPages());
    map.put("currentPage", page+1);
    return "index";
}
```

接口中除了向map中放置博客列表的数据外，还放置了分页相关的数据，比如总页数、当前页码等。

博客列表页面也是系统主页，所以命名为index.html，与接口返回值对应，并将index.html放置在目录template下。在index.html中，只需要获取blogs、total、currentPage等变量并显示出来即可，具体代码如下：

```html
<!DOCTYPE html>
<html lang="en" xmlns:th="https://www.thymeleaf.org">
<head>
    <meta charset="UTF-8">
    <title>博客列表</title>
    <style>
        table,table tr th, table tr td { border:1px solid #000000; }
    </style>
</head>
<body>
```

```html
<h2>个人博客系统-博客列表</h2>
<table style="border:1px solid #ccc" rules="rows">
  <thead>
    <tr>
      <th>标题</th>
      <th>创建时间</th>
      <th>浏览次数</th>
      <th>喜欢次数</th>
    </tr>
  </thead>
  <tbody id="list">
    <tr th:each="blog:${blogs}">
      <td th:text="${blog.title}"></td>
      <td th:text="${blog.publishTime}"></td>
      <td th:text="${blog.scanCount}"></td>
      <td th:text="${blog.likeCount}"></td>
    </tr>
  </tbody>
  <tfoot>
    <tr>
      <td>
        <a th:href="@{'/blog?page=' + ${pageLast} + '&size=' + ${size}}">上一页</a>
      </td>
      <td>
        <a th:href="@{'/blog?page=' + ${pageNext} + '&size=' + ${size}}">下一页</a>
      </td>
      <td>当前第<span th:text="${currentPage}"></span>页</td>
      <td>总共<span th:text="${total}"></span>页</td>
    </tr>
  </tfoot>
</table>
</body>
</html>
```

然后重启/启动项目，在浏览器中访问地址http://127.0.0.1:8080/blog，效果如图11.17所示。

图11.17 博客列表页面

单击"下一页"链接，跳转到第二页，页面效果如图11.18所示。

图 11.18 博客列表页面第二页

单击"上一页"链接，又可以返回首页。到这里便完成了博客列表的开发，只是现在还不支持打开一篇博客，到下一节中再来修改，以支持打开博客详情页面。

11.5.3 开发博客详情接口和页面

暂不考虑博客喜欢接口和浏览次数接口，博客详情页面只是加载博客内容并展示在页面上，所以非常适合使用模板引擎来实现。

在BlogController类下定义blog方法，参数为int类型，用于接口查询博客的id。使用注解@GetMapping，并标识为接口GET请求。在方法中根据id查询博客信息，然后将信息设置到map中，代码如下：

```
@GetMapping("/{id}")
public String blog(@PathVariable int id, Map<String, Object> map){
    Optional<Blog> blogOpt = blogRepository.findById(id);
    if(blogOpt.isEmpty()){
        return "error";
    }
    Blog blog = blogOpt.get();
    map.put("id", blog.getId());
    map.put("title", blog.getTitle());
    map.put("publishTime", new SimpleDateFormat("yyyy年MM月dd日 HH时mm分ss秒").format(blog.getPublishTime()));
    map.put("scanCount", blog.getScanCount());
    map.put("likeCount", blog.getLikeCount());
    map.put("content", blog.getContent());
    return "blog";
}
```

在目录template下创建blog.html文件，页面逻辑主要是将接口设置到map中的数据取出来，具体代码如下：

```
<!DOCTYPE html>
<html lang="en" xmlns:th="https://www.thymeleaf.org">
<head>
  <meta charset="UTF-8">
  <title></title>
  <script src="/js/jquery-3.6.0.min.js"></script>
```

```
</head>
<body>
<h2 th:text="${title}"></h2>
<hr>
<table>
  <tr>
    <td>发布时间：<span th:text="${publishTime}"></span></td>
    <td width="20px"></td>
    <td>浏览次数：<span th:text="${scanCount}" id="scan_count"></span></td>
    <td width="20px"></td>
    <td>喜欢次数：<span th:text="${likeCount}"></span></td>
  </tr>
</table>
<hr>
<div th:text="${content}">
</div>
<script>
</script>
</body>
</html>
```

启动项目，在浏览器中访问地址 http://127.0.0.1:8080/blog/1，页面响应如图11.19所示。

图 11.19　博客详情页面

当前博客详情页面已经完成，但是还没有配置页面的入口。下面修改博客列表页面，将博客标题修改为链接，单击博客标题打开博客详情页。将index.html中的博客标题td，也就是下面的代码：

```
<td th:text="${blog.title}"></td>
```

修改为链接<a>的格式，具体代码如下：

```
<td>
    <a th:text="${blog.title}" th:href="@{'/blog/'+${blog.id}}"></a>
</td>
```

再访问博客列表页，此时标题列已经修改成了链接，如图11.20所示。单击标题链接，页面跳转到博客详情页面。

图 11.20　博客列表页面增加链接

11.5.4 实现浏览次数计数功能

浏览次数计数功能实现博客详情页每被打开一次，浏览次数增加一。实现的思路是在"浏览次数计数"接口用来将浏览次数加一，博客详情页面每次打开时通过Ajax调用"浏览次数计数"接口。

开发"浏览次数计数"接口，在BlogController中增加scan方法，使用注解@ResponseBody标注，表示此接口返回内容而不是页面。接口参数为博客id，然后调用blogRepository的addScanCount方法将指定博客的scan_count加一，addScanCount方法定义如下：

```
@Transactional
@Modifying
@Query(nativeQuery = true,
        value = "update blogs set scan_count = scan_count+1 where id = ?1")
int addScanCount(int id);
```

因为涉及数据修改，所以要使用注解@Transactional和注解@Modifying。接口方法scan在更新scan_count字段之后，查询字段的最新值并返回，代码如下：

```
@PostMapping("/{id}/scan")
@ResponseBody
public int scan(@PathVariable int id){
    blogRepository.addScanCount(id);
    return blogRepository.getById(id).getScanCount();
}
```

然后修改日志详情页面，调用接口时需要传递id，为了方便获取id，在blog.html中增加一个隐藏的id文本框，代码如下：

```
<input type="text" th:value="${id}" id="blog_id" style="display:none"></input>
```

id的值在博客详情接口中已经设置，所以上面的代码可以取到。然后在blog.html中添加访问"浏览次数计数"接口的JS代码，代码如下：

```
let blogId = $("#blog_id").val();
alert(blogId);
$.ajax({
    type: 'POST',
    url: '/blog/' + blogId + '/scan',
    contentType: 'application/json',
    withCredentials: true,
    success: function(data){
        $("#scan_count").text(data);
    },
    error: function(data){
        // 失败后执行的代码
        alert("浏览计数失败！");
        console.log(data);
        return;
    }
});
```

当接口成功响应时，将最新的浏览次数设置到浏览次数的span标签中。

完成代码后重启/启动项目，在浏览器中访问地址http://127.0.0.1:8080/blog/1，可以看到浏览次数修改成了1，然后刷新页面，可以看到浏览次数增加1，显示为2，如图11.21所示。

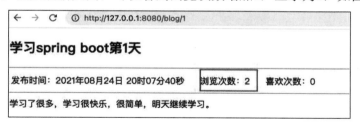

图 11.21　浏览次数增加

11.6　开发用户模块

本节将要完成用户模块相关的功能，有默认的管理员用户配置、用户登录、喜欢的接口配置以及对发布博客的接口和页面限制仅允许管理员操作。登录页面和接口相对独立，下面先完成接口的开发，再来开发页面。

11.6.1　开发登录相关接口

1. 发送验证码接口

手机号发送验证码的大概逻辑是：接口接收手机号，然后生成随机验证码，调用发送短信的服务将验证码发送出去，然后以手机号为key保存验证码，可以保存到内存、数据库或者Redis等。

在示例项目中不涉及真实的发送验证码服务，所以使用Map存储在内存中。在验证码验证通过时，将存储的验证码移除。为了代码的可复用性，将生成验证码和校验验证码的代码封装在一个类中，命名为SmsService，代码如下：

```java
package com.example.selfblog.repository;

import org.springframework.stereotype.Component;

import java.util.HashMap;
import java.util.Map;

@Component
public class SmsService {
    /**
     * 模拟存储，在实际项目中可以用数据库或者Redis等来存储
     */
    private static Map<String, String> vcMap = new HashMap<>();

    /**
     * 生成验证码，并以手机号为key存储起来，然后返回生成的验证码
```

```java
    */
    public String genVerifyCode(String phoneNumber) {
        String vc = "123456"; // 这里为了测试方便,写为123456。
        vcMap.put(phoneNumber, vc);
        return vc;
    }

    /**
     * 校验用户传递的验证码是否正确。返回true表示匹配,false表示不匹配
     */
    public boolean checkVerifyCode(String phoneNumber, String verifyCode) {
        String vc = vcMap.get(phoneNumber);
        if (vc == null) {
            return false;
        } else {
            if(vc.equals(verifyCode)){
                vcMap.remove(phoneNumber);
                return true;
            }else {
                return false;
            }
        }
    }
}
```

接下来定义发送验证码的HTTP接口,接口实现逻辑相对简单,只需要调用genVerifyCode即可完成。创建LoginController类,使用@RestController和@RequestMapping("login")标注,然后定义接口方法sendSms,定义为接收PUT请求,发送成功时返回1,代码如下:

```java
package com.example.selfblog.controller;

import com.example.selfblog.pojo.UserLogin;
import com.example.selfblog.repository.SmsService;
import org.springframework.beans.factory.annotation.Autowired;
import org.springframework.web.bind.annotation.*;

@RestController
@RequestMapping("login")
public class LoginController {
    @Autowired
    SmsService smsService;

    @PutMapping("sms")
    public String sendSms(@RequestBody UserLogin userLogin){
        String verifyCode =
smsService.genVerifyCode(userLogin.getPhoneNumber());
        return "1";
    }
}
```

接口方法的形参为UserLogin类型,UserLogin会被反复用到,具体代码如下:

```
package com.example.selfblog.pojo;

import lombok.AllArgsConstructor;
import lombok.Data;
import lombok.NoArgsConstructor;

@Data
@AllArgsConstructor
@NoArgsConstructor
public class UserLogin {
    private String phoneNumber;
    private String verifyCode;
    private String password;
}
```

2. 验证码登录接口

接收浏览器传过来的手机号和验证码，然后根据内存中的数据做校验，验证通过返回token，验证不通过返回0，代码如下：

```
@PostMapping("byvc")
public String loginByVerifyCode(@RequestBody UserLogin userLogin){
    boolean flag = smsService.checkVerifyCode(userLogin.getPhoneNumber(), userLogin.getVerifyCode());
    if(flag){
        return loginService.loginOrRegister(userLogin.getPhoneNumber());
    }else{
        return "0";
    }
}
```

验证码通过验证后，如果是首次登录，还需要将用户信息存储到数据库，上面在数据库中创建用户的过程封装在了loginService.loginOrRegister方法中。由于需要访问数据库，因此使用JPA Repository定义UserRepository接口，接口定义代码如下：

```
package com.example.selfblog.repository;

import com.example.selfblog.entity.User;
import org.springframework.data.jpa.repository.JpaRepository;
import org.springframework.data.jpa.repository.JpaSpecificationExecutor;

public interface UserRepository extends JpaRepository<User, Integer>, JpaSpecificationExecutor<User> {
}
```

创建LoginService类，并实现loginOrRegister方法，具体代码如下：

```
package com.example.selfblog.service.impl;

import com.example.selfblog.config.security.TokenService;
import com.example.selfblog.entity.User;
import com.example.selfblog.repository.UserRepository;
import com.example.selfblog.service.LoginService;
import org.springframework.beans.factory.annotation.Autowired;
```

```java
import org.springframework.security.authentication.UsernamePasswordAuthenticationToken;
import org.springframework.security.core.context.SecurityContextHolder;
import org.springframework.security.crypto.password.PasswordEncoder;
import org.springframework.stereotype.Service;

import java.sql.Timestamp;

@Service
public class LoginServiceImpl implements LoginService {
    @Autowired
    PasswordEncoder passwordEncoder;
    @Autowired
    TokenService tokenService;
    @Autowired
    UserRepository userRepository;

    @Override
    public String loginOrRegister(String phoneNumber) {
        User user = userRepository.findOneByPhoneNumber(phoneNumber);
        if (null == user) { // 用户不存在，需要先创建用户
            user = new User();
            user.setPhoneNumber(phoneNumber);
            user.setPassword("");
            long currentTimeStamp = System.currentTimeMillis();
            user.setUpdateTime(new Timestamp(currentTimeStamp));
            userRepository.save(user);
        }
        return updateSecurityUserInfo(user);
    }

    public String updateSecurityUserInfo(User user) {
        // 更新 Security 登录用户的信息
        UsernamePasswordAuthenticationToken usernamePasswordAuthenticationToken
                = new UsernamePasswordAuthenticationToken(user, null, null);
        SecurityContextHolder.getContext().setAuthentication(usernamePasswordAuthenticationToken);

        String token = tokenService.generateToken(user);
        return token;
    }
}
```

上面使用了TokenService类，这个类是我们对JWT的一次封装，为了使用JWT，需要先添加JWT依赖，在pom.xml中添加如下配置：

```xml
<dependency>
    <groupId>io.jsonwebtoken</groupId>
    <artifactId>jjwt</artifactId>
    <version>0.9.1</version>
</dependency>
```

TokenService类代码如下：

```java
package com.example.selfblog.config.security;

import com.example.selfblog.entity.User;
import io.jsonwebtoken.Claims;
import io.jsonwebtoken.Jwts;
import io.jsonwebtoken.SignatureAlgorithm;
import org.springframework.stereotype.Component;

import java.util.Date;
import java.util.HashMap;
import java.util.Map;

@Component
public class TokenService {
    private static final String SUB = "sub";

    /**
     * 根据用户信息生成 Token
     */
    public String generateToken(User user) {
        Map<String, Object> claims = new HashMap<>();
        claims.put(SUB, user.getId());
        return generateToken(claims);
    }

    private String generateToken(Map<String, Object> claims) {
        return Jwts.builder().setClaims(claims).setExpiration(new Date(System.currentTimeMillis() + 7 * 24 * 3600000L))
                .signWith(SignatureAlgorithm.HS256, "secret").compact();
    }

    /**
     * 从 token 中获取用户 id
     */
    public Integer getUserIdFromToken(String token) {
        String userId;
        try {
            Claims claims = getClaimsFromToken(token);
            userId = claims.getSubject();
        } catch (Exception e) {
            return null;
        }
        return Integer.valueOf(userId);
    }

    /**
     * 从 token 中获取 Claim
     */
    private Claims getClaimsFromToken(String token) {
        Claims claims = null;
        try {
            claims = Jwts.parser().setSigningKey("secret")
                    .parseClaimsJws(token)
                    .getBody();
```

```java
        } catch (Exception e) {
            e.printStackTrace();
        }
        return claims;
    }

    /**
     * 判断 token 是否有效 && 是否过期
     */
    public boolean validateToken(String token, User user) {
        // 判断是否过期, 判断是否有效
        Integer userId = getUserIdFromToken(token);
        return userId.equals(user.getId())
                && !isTokenExpired(token);
    }

    public boolean isTokenExpired(String token) {
        Date expireDate = getExpireDateFromToken(token);
        return expireDate.before(new Date());
    }

    public Date getExpireDateFromToken(String token) {
        Claims claims = getClaimsFromToken(token);
        return claims.getExpiration();
    }
}
```

至此，便完成了验证码登录接口。

3. 修改密码接口

用户登录后才可以修改密码，并且修改密码前需要发送验证码，根据前台传递的验证码与session中解析出的手机号进行验证，如果验证通过，就对密码进行加密处理，加密后的字段存储到数据库的password字段中。

关于密码加密，这里使用BCryptPasswordEncoder，为了在容器中定义Bean，在Security的Config类中添加如下代码：

```java
@Bean
PasswordEncoder passwordEncoder(){
    return new BCryptPasswordEncoder();
}
```

创建UserController，并在其中注入PasswordEncoder和UserRepository，然后定义修改密码的HTTP接口方法setPasswordByVerifyCode，方法代码如下：

```java
@ResponseBody
@PostMapping("chpw")
public String setPasswordByVerifyCode(@RequestBody UserLogin userLogin) {
    User user = UserUtil.getUser();
    if (smsService.checkVerifyCode(user.getPhoneNumber(),
userLogin.getVerifyCode())) {
        user.setPassword(passwordEncoder.encode(userLogin.getPassword()));
```

```
            userRepository.save(user);
            return "1";
        } else {
            return "0";
        }
    }
```

4. 密码登录接口

密码登录就是对用户录入的手机号和密码与根据手机号从数据库中查询出来的密码进行校验，如果校验通过，就返回token。

在LoginService接口中定义根据密码登录的方法loginByPW，并在类LoginServiceImpl中完成实现，方法代码如下：

```
@Override
public String loginByPW(String phoneNumber, String password) {
    User user = userRepository.findOneByPhoneNumber(phoneNumber);
    if (null == user || !passwordEncoder.matches(password, user.getPassword())) {
        return "0";
    }
    return updateSecurityUserInfo(user);
}
```

回到LoginController类中，定义方法loginByPassword，并在接口参数后调用LoginService的loginByPW方法，代码如下：

```
@PostMapping("bypw")
public String loginByPassword(@RequestBody UserLogin userLogin){
    return loginService.loginByPW(userLogin.getPhoneNumber(),
userLogin.getPassword());
}
```

11.6.2 完成登录页面

登录页面初始化不需要加载数据，所以可以使用HTML来实现页面，通过jQuery调用Ajax来实现登录操作。

在static目录下创建login.html，页面中需要验证码登录和密码登录两部分，两部分通过水平线标签<hr>分割。页面中涉及3个按钮，一个发送验证码按钮和两个登录按钮。发送验证码按钮被单击时调用发送验证码接口，两个登录按钮分别用于调用各自的登录接口。login.html具体代码如下：

```
<!DOCTYPE html>
<body>
<h1>个人博客系统-登录</h1>
<hr>
<h3>使用验证码登录</h3>
<div>
    <table>
        <tr>
```

```html
            <td>手机号</td>
            <td><input id="phoneNumber" type="text"></td>
        </tr>
        <tr>
            <td colspan="2"><button onclick="sendSms()">发送验证码</button></td>
        </tr>
        <tr>
            <td>验证码</td>
            <td><input id="verifyCode" type="text"></td>
        </tr>
        <tr>
            <td colspan="2">
                <button onclick="loginByVC(this)">保存</button>
            </td>
        </tr>
    </table>
</div>
<hr>
<h3>使用密码登录</h3>
<div>
    <table>
        <tr>
            <td>手机号</td>
            <td><input id="phoneNumber1" type="text"></td>
        </tr>
        <tr>
            <td>密码</td>
            <td><input id="password1" type="password"></td>
        </tr>
        <tr>
            <td colspan="2">
                <button onclick="loginByPW(this)">保存</button>
            </td>
        </tr>
    </table>
</div>
<script>
function sendSms(){
    var phoneNumber = $("#phoneNumber").val();
    var param_data = {"phoneNumber" : phoneNumber };
    $.ajax({
        type: 'PUT',
        url: '/login/sms',
        contentType: 'application/json',
        withCredentials: true,
        data: JSON.stringify(param_data),
        success: function(data){
            alert("发送验证码成功！");
        },
```

```javascript
        error: function(data){
            // 失败后执行的代码
            alert("发送验证码失败！");
            console.log(data);
            return;
        }
    });
}
function loginByVC(){
    var phoneNumber = $("#phoneNumber").val();
    var verifyCode = $("#verifyCode").val();
    var param_data = {"phoneNumber" : phoneNumber, "verifyCode": verifyCode};
    $.ajax({
        type: 'POST',
        url: '/login/byvc',
        contentType: 'application/json',
        withCredentials: true,
        data: JSON.stringify(param_data),
        success: function(data){
            if("0"==data){
                alert("登录失败！");
            }else{
                sessionStorage.setItem("token", data);
                alert("登录成功！");
                window.location.href = "/";
            }
        },
        error: function(data){
            // 失败后执行的代码
            alert("登录失败！");
            console.log(data);
            return;
        }
    });
}
function loginByPW(){
    var phoneNumber = $("#phoneNumber1").val();
    var password = $("#password1").val();
    var param_data = {"phoneNumber" : phoneNumber, "password" : password };
    $.ajax({
        type: 'POST',
        url: '/login/bypw',
        contentType: 'application/json',
        withCredentials: true,
        data: JSON.stringify(param_data),
        success: function(data){
            if("0"==data){
                alert("登录失败！");
            }else{
```

```
                sessionStorage.setItem("token", data);
                alert("登录成功！");
                window.location.href = "/";
            }
        },
        error: function(data){
            // 失败后执行的代码
            alert("登录失败！");
            console.log(data);
            return;
        }
    });
}
</script>
</body>
</html>
```

11.6.3 测试用户登录功能

启动/重启项目，在浏览器中访问地址http://127.0.0.1:8080/login.html，效果如图11.22所示。

首先要录入手机号，单击"发送验证码"按钮，浏览器提示发送成功，提示框如图11.23所示。

图 11.22 登录页面

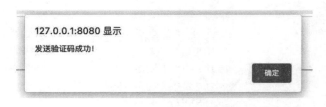

图 11.23 发送验证码提示框

由于我们写死的验证码是"123456"，所以在验证码文本框中录入"123456"，如图11.24所示。

然后单击验证码登录部分的"登录"按钮，浏览器提示登录成功，如图11.25所示。

图 11.24 登录信息

图 11.25 登录成功提示框

单击"确定"按钮,浏览器跳转到博客列表页面。

11.6.4 在博客列表页面增加当前用户的显示

由于非登录用户也可以访问博客列表页面,因此对用户来讲,在博客列表页面不容易区分是否已登录。为此,在博客列表页面再加一部分内容,对于登录的用户显示用户的手机号,单击用户手机号跳转到用户首页,用户首页内置修改密码的功能。对于未登录的用户,显示一个"登录"链接,单击链接跳转到登录页面。

1. 在列表页面接口放置登录用户的信息

为了在接口中方便地获取登录用户的信息,先定义一个Filter,用来从token中解析用户信息,并将解析后的信息放置到上下文中。在config.security包下创建类TokenFilter,继承OncePerRequestFilter并重写doFilterInternal方法,代码如下:

```
package com.example.selfblog.config.security;

import com.example.selfblog.entity.User;
import com.example.selfblog.repository.UserRepository;
import lombok.extern.slf4j.Slf4j;
import org.springframework.beans.factory.annotation.Autowired;
import org.springframework.orm.jpa.JpaObjectRetrievalFailureException;
import org.springframework.security.authentication.
UsernamePasswordAuthenticationToken;
import org.springframework.security.core.context.SecurityContextHolder;
import org.springframework.security.web.authentication.
WebAuthenticationDetailsSource;
import org.springframework.web.filter.OncePerRequestFilter;

import javax.servlet.FilterChain;
import javax.servlet.ServletException;
import javax.servlet.http.HttpServletRequest;
import javax.servlet.http.HttpServletResponse;
import java.io.IOException;

@Slf4j
public class TokenFilter extends OncePerRequestFilter {
    @Autowired
    TokenService tokenService;
    @Autowired
    UserRepository userRepository;

    @Override
    protected void doFilterInternal(HttpServletRequest request,
HttpServletResponse response, FilterChain filterChain) throws ServletException,
IOException {
        String token = request.getHeader("token");
        if(token!=null && !"".equals(token.trim())){//
&& !token.startsWith("JSESSIONID")){
            Integer userId = tokenService.getUserIdFromToken(token);
            // 可以从token中拿到用户名,但是未登录
```

```java
                if (null != userId && null == 
SecurityContextHolder.getContext().getAuthentication()) {
                User user = null;
                try {
                    user = userRepository.getById(userId);
                } catch (JpaObjectRetrievalFailureException e) {
                    log.error("token 中的userId 无效" + userId, e);
                    throw new RuntimeException("token 中的userId无效" + userId);
                }
                if (user != null && tokenService.validateToken(token, user)) {
                    UsernamePasswordAuthenticationToken 
usernamePasswordAuthenticationToken = new 
UsernamePasswordAuthenticationToken(user, null, null);
                    usernamePasswordAuthenticationToken.setDetails(new 
WebAuthenticationDetailsSource().buildDetails(request));
                    SecurityContextHolder.getContext().setAuthentication
(usernamePasswordAuthenticationToken);
                } else {
                    log.error("登录状态失效，请重新登录。");
                    throw new RuntimeException("登录状态失效，请重新登录。");
                }
            }
        }
        filterChain.doFilter(request, response);
    }
}
```

然后定义一个User工具类UserUtil，从SecurityContextHolder中获取用户信息，UserUtil代码如下：

```java
package com.example.selfblog.util;

import com.example.selfblog.entity.User;
import org.springframework.security.core.Authentication;
import org.springframework.security.core.context.SecurityContextHolder;

public class UserUtil {
    public static User getUser(){
        Authentication authentication = 
SecurityContextHolder.getContext().getAuthentication();
        if(authentication == null){
            return null;
        }
        try{
            return (User) 
SecurityContextHolder.getContext().getAuthentication().getPrincipal();
        }catch (Exception e){
            return null;
        }
    }
}
```

如此一来，只要在需要用户信息的时候，调用UserUtil的getUser方法便可以获取到。

再到BlogController类中，为了将HTTP请求根目录设置到博客列表页面，先将类上的@RequestMapping注解去掉，把方法上的@RequestMapping配置的路径统一加上前缀"blog/"，然后修改列表方法blogs的路径，修改后如下：

```
@GetMapping({"blog",""})
```

在blogs方法返回前，为了给页面返回用户信息，在map中设置user属性，具体代码如下：

```
map.put("user", UserUtil.getUser());
```

2. 列表页面显示登录用户的信息或者登录链接

在列表页面需要判断是否为登录用户，如果是登录用户，则显示"当前用户："拼接上手机号；如果不是登录用户，则显示"未登录，去登录"。

单击登录用户的链接打开用户主页，单击未登录用户的链接打开登录页面。

在博客列表页面index.html的</table>和</body>之间增加如下代码：

```
<hr>
<a href="/user" th:text="${user !=null ? '当前用户：'+ user.phoneNumber : ''}"></a>
<a href="/login.html" th:text="${user !=null ? '' : '未登录，去登录'}"></a>
<hr>
```

然后重启项目，访问系统根目录地址http://127.0.0.1:8080/，效果如图11.26所示。

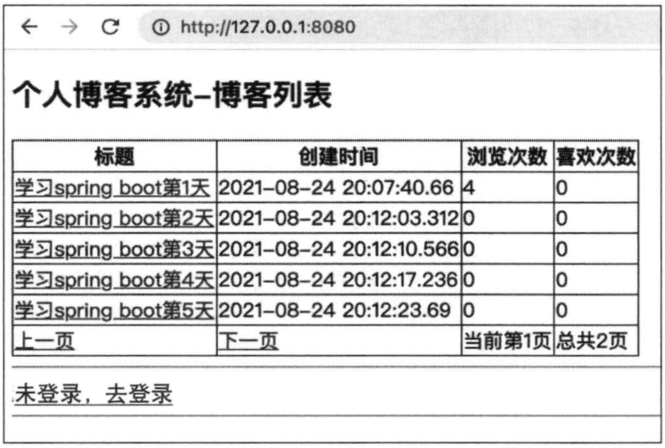

图11.26　博客列表增加登录信息

单击底部的"未登录，去登录"链接，页面跳转到登录页面，如图11.27所示。

录入手机号"1520000111*"，单击"发送验证码"按钮，提示发送验证码成功，单击"确认"按钮，然后录入验证码"123456"，如图11.28所示。

单击上面的"登录"按钮，页面提示"登录成功"，单击"确认"按钮，页面跳转到博客列表页，页面底部显示当前用户信息，如图11.29所示。

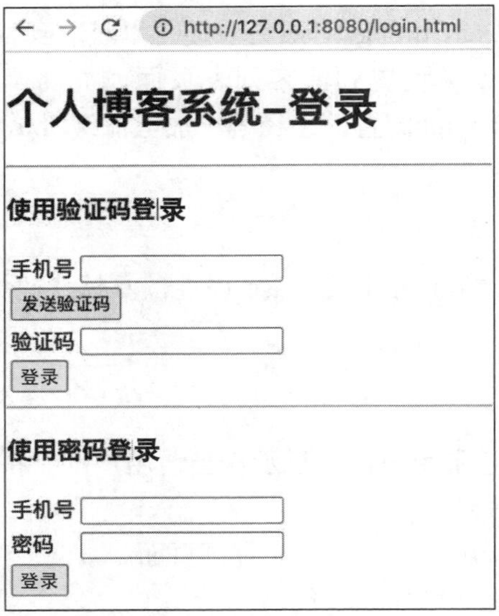

图 11.27 跳转到登录页面 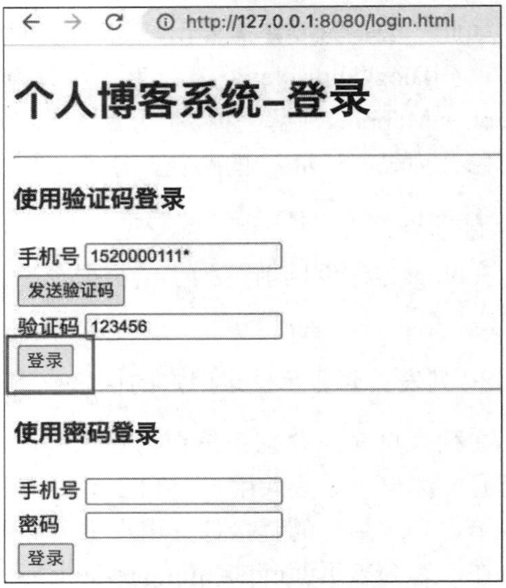　　　　　　　　　图 11.28 录入验证码登录信息

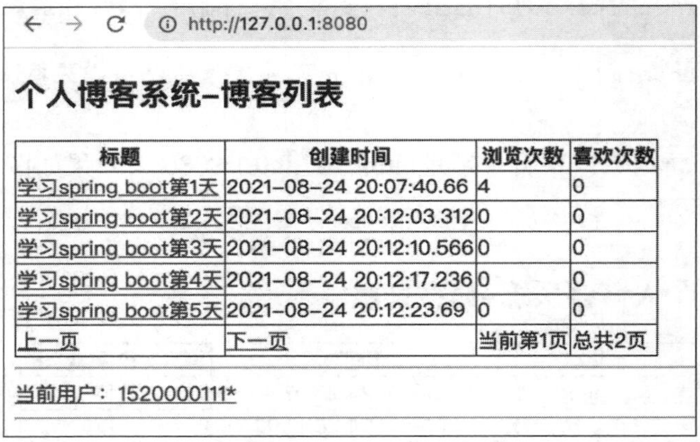

图 11.29 博客列表页面显示当前用户

11.6.5　个人主页页面

个人主页分成两部分，上面部分显示用户的 id 和用户手机号，下面显示修改密码的表格。页面使用 Thymleaf 模板引擎来实现。

1. 个人主页接口

在 UserController 中创建 userIndex，在方法中获取当前登录用户的信息，并以 user 为 key 设置到 map 中，具体代码如下：

```
@GetMapping
public String userIndex(Map<String, Object> map){
    map.put("user", UserUtil.getUser());
    return "self_page";
}
```

2. 个人主页页面

在template目录下创建self_page.html，页面结构简单，只是在单击"发送验证码"按钮和"修改密码"按钮的时候，需要触发JS方法，具体代码如下：

```
<!DOCTYPE html>
<body>
<h2>个人博客系统-个人首页</h2>
<span th:text="${user !=null ? '用户ID: '+ user.id : ''}"></span>
<span th:text="${user !=null ? '用户手机号: '+ user.phoneNumber : ''}"></span>
<hr>
<h4>修改密码</h4>
<div>
  <table>
    <tr>
      <td colspan="2"><button onclick="sendSms()">发送验证码</button></td>
    </tr>
    <tr>
      <td>验证码</td>
      <td><input id="verifyCode" type="text"></td>
    </tr>
    <tr>
      <td>设置密码</td>
      <td><input id="password" type="password"></td>
    </tr>
    <tr>
      <td colspan="2">
        <button onclick="resetPwd(this)">修改密码</button>
      </td>
    </tr>
  </table>
</div>
<hr>
<input id="phoneNumber" th:value="${user !=null ? user.phoneNumber : ''}"/>
<script>
function sendSms(){
    var phoneNumber = $("#phoneNumber").val();
    var param_data = {"phoneNumber" : phoneNumber };
    $.ajax({
        type: 'PUT',
        url: '/login/sms',
        contentType: 'application/json',
        withCredentials: true,
        data: JSON.stringify(param_data),
        success: function(data){
            alert("发送验证码成功！");
        },
        error: function(data){
            // 失败后执行的代码
            alert("发送验证码失败！");
```

```
                console.log(data);
                return;
            }
        });
    }
    function resetPwd(){
        var verifyCode = $("#verifyCode").val();
        var password = $("#password").val();
        var param_data = {"verifyCode" : verifyCode, "password" : password };
        $.ajax({
            type: 'POST',
            url: '/user/chpw',
            contentType: 'application/json',
            withCredentials: true,
            data: JSON.stringify(param_data),
            success: function(data){
                if("1"==data){
                    alert("修改成功！");
                }else{
                    alert("修改失败！");
                }
            },
            error: function(data){
                // 失败后执行的代码
                alert("登录失败！");
                console.log(data);
                return;
            }
        });
    }
</script>
</body>
</html>
```

启动/重启项目，在浏览器访问地址http://127.0.0.1:8080/，重启后需要先登录，单击"未登录，去登录"链接进入登录页面，然后仍然以手机号"1520000111*"和验证码来登录。登录后，浏览器跳转到博客列表页，单击底部的链接"当前用户：1520000111*"，浏览器跳转到个人首页时，页面如图11.30所示。

单击"发送验证码"按钮，提示发送成功后，单击"确定"按钮，然后录入验证"123456"，设置密码为"123abc"，如图11.31所示。

单击"修改密码"按钮，提示"修改成功"。由于没有做退出登录功能，可以重启项目来使认证失效。重启项目后访问登录页面（http://127.0.0.1:8080/login.html），再使用密码登录模块录入手机号和密码，并录入修改的密码，如图11.32所示。

图 11.30　个人首页页面

图 11.31　录入修改密码信息

图 11.32　录入密码登录信息

然后单击下面的"登录"按钮，浏览器提示登录成功，单击"确定"按钮后，页面跳转到博客列表页，页面底部显示当前登录用户的信息，如图11.33所示，完成修改密码功能。

图 11.33　使用密码登录后进入博客列表页面

11.7　实现喜欢/取消喜欢博客功能

在博客详情页，对于登录的用户，应该有"喜欢"和"取消喜欢"按钮，对于没有登录的用户，不显示这两个按钮。

"喜欢"和"取消喜欢"按钮被单击时，分别调用"喜欢博客"接口和"取消喜欢博客"接口。这两个按钮/接口都会触发博客喜欢数量的变动，在单击按钮触发接口被成功调用后，将博客详情页的喜欢数量加一或者减一。

11.7.1　开发"喜欢博客"接口

喜欢博客接口从请求路径中接收博客id作为参数，然后查询当前用户和博客信息，并做非空校验，封装到UserLikedBlog实体类中，然后保存到数据库，并对博客的like_count字段加一。

在BlogRepository中定义对博客like_count字段加一的方法，代码如下：

```
@Transactional
@Modifying
@Query(nativeQuery = true,
       value = "update blogs set like_count = like_count+1 where id = ?1")
int addLikeCount(int id);
```

创建BlogRepository的repository接口UserLikedBlogRepository，在repository接口下创建接口，代码如下：

```
package com.example.selfblog.repository;

import com.example.selfblog.entity.UserLikedBlog;
import org.springframework.data.jpa.repository.JpaRepository;
import org.springframework.data.jpa.repository.JpaSpecificationExecutor;
public interface UserLikedBlogRepository extends JpaRepository<UserLikedBlog, Integer>, JpaSpecificationExecutor<UserLikedBlog> {
}
```

然后在**BlogController**中定义**like**方法作为HTTP接口方法，具体代码如下：

```
@Transactional
@PostMapping("blog/{id}/like")
@ResponseBody
public int like(@PathVariable int id) {
    User user = UserUtil.getUser();
    if (user == null) {
        return 0;
    }
    Blog blog = blogRepository.getById(id);
    if (blog == null) {
        return 0;
    }
    UserLikedBlog userLikedBlog = new UserLikedBlog();
    userLikedBlog.setUser(user);
    userLikedBlog.setBlog(blog);
    userLikedBlog.setLikeTime(new Timestamp(System.currentTimeMillis()));
    userLikedBlogRepository.save(userLikedBlog);
    blogRepository.addLikeCount(id);
    return 1;
}
```

这里因为是对两张表进行操作，所以使用了@Transactional注解。

11.7.2 开发"取消喜欢博客"接口

"取消喜欢博客"接口从请求路径上获取博客id，并在方法内部获取当前用户信息。根据博客id和用户id从user_liked_blog删除记录，然后将博客表对应的数据的like_count减一。

在BlogRepository中定义方法subLikeCount，用来实现将like_count减一的操作，代码如下：

```
@Transactional
@Modifying
@Query(nativeQuery = true,
       value = "update blogs set like_count = like_count-1 where id = ?1")
int subLikeCount(int id);
```

在UserLikedBlogRepository中定义根据博客id和用户id删除喜欢博客记录的方法delUserLikedBlogs，代码如下：

```
@Transactional
@Modifying
@Query(nativeQuery = true,
       value = "delete from user_liked_blog where blog_id=?1 and user_id=?2 ")
int delUserLikedBlogs(int blogId, int userId);
```

然后在BlogController中定义dislike方法作为HTTP接口方法，具体代码如下：

```
@Transactional
@PostMapping("blog/{id}/dislike")
@ResponseBody
public int dislike(@PathVariable int id) {
    User user = UserUtil.getUser();
    if (user == null) {
        return 0;
    }
    userLikedBlogRepository.delUserLikedBlogs(id, user.getId());
    blogRepository.subLikeCount(id);
    return 1;
}
```

11.7.3　修改博客详情页面接口，返回当前用户是否已喜欢

在UserLikedBlogRepository接口中，使用关键字countBy来定义方法，查询当前用户是否有喜欢的记录，方法定义如下：

```
int countByBlogIdAndUserId(int blogId, int userId);
```

然后回到BlogController类中的blogs方法，获取当前登录用户的id，结合博客id查询是否有喜欢记录，如果有，则在map中设置标识字段hasLiked为1，否则设置为0。如果用户没有登录，则不设置标识字段。修改后的blog方法的代码如下：

```
@GetMapping("blog/{id}")
public String blog(@PathVariable int id, Map<String, Object> map) {
    Optional<Blog> blogOpt = blogRepository.findById(id);
    if (blogOpt.isEmpty()) {
        return "error";
    }
    Blog blog = blogOpt.get();
    map.put("id", blog.getId());
```

```
            map.put("title", blog.getTitle());
            map.put("publishTime", new SimpleDateFormat("yyyy年MM月dd日 HH时mm分ss
秒").format(blog.getPublishTime()));
            map.put("scanCount", blog.getScanCount());
            map.put("likeCount", blog.getLikeCount());
            map.put("content", blog.getContent());
            User user = UserUtil.getUser();
            if (user != null) {
                int hasLiked = userLikedBlogRepository.countByBlogIdAndUserId(id,
user.getId());
                if (hasLiked > 0) {
                    map.put("hasLiked", "1");
                } else {
                    map.put("hasLiked", "0");
                }
            }
            return "blog";
    }
```

11.7.4 修改博客详情页面，增加喜欢/取消喜欢按钮

在博客详情页面上，根据hasLiked字段来判断是否显示喜欢博客或者取消喜欢博客按钮。在单击按钮后，如果接口响应为1，还需要对当前页面显示的喜欢数量做加一或减一的处理，以及需要将"喜欢按钮"修改为"取消喜欢按钮"，或者将"取消喜欢"按钮修改为"喜欢"按钮。

通过Thymeleaf属性可以实现按钮的显示或者隐藏，在blog.html中喜欢次数td的后面增加两个td，具体代码如下：

```
<td width="20px"></td>
<td th:if="${hasLiked} != null" id="btnTd">
  <button th:if="${hasLiked} eq '1'" onclick="dislike()">取消喜欢</button>
  <button th:if="${hasLiked} eq '0'" onclick="like()">喜欢</button>
</td>
```

然后在JS代码中增加dislike和like两个方法，JS代码如下：

```
function like(){
    $.ajax({
        type: 'POST',
        url: '/blog/' + blogId + '/like',
        withCredentials: true,
        success: function(data){
            if("1" == data){
                let likeCountSpan = $("#likeCount");
                likeCountSpan.text( parseInt(likeCountSpan.text()) + 1 );
                $("#btnTd").html('<button onclick="dislike()">取消喜欢</button>');
            }
        },
        error: function(data){
```

```
            // 失败后执行的代码
            alert("发布评论失败！如未登录，请先登录！");
            console.log(data);
            return;
        }
    });
}
function dislike(){
    $.ajax({
        type: 'POST',
        url: '/blog/' + blogId + '/dislike',
        withCredentials: true,
        success: function(data){
            if("1" == data){
                let likeCountSpan = $("#likeCount");
                likeCountSpan.text( parseInt(likeCountSpan.text()) - 1 );
                $("#btnTd").html('<button onclick="like()">喜欢</button>');
            }
        },
        error: function(data){
            // 失败后执行的代码
            alert("发布评论失败！如未登录，请先登录！");
            console.log(data);
            return;
        }
    });
}
```

11.7.5　页面测试

启动项目，在浏览器中访问地址http://127.0.0.1:8080，打开博客列表页面，如图11.34所示。

图 11.34　未登录时的博客列表页面

可以看到当前是未登录的状态，然后单击第一篇博客，详情页面显示如图11.35所示。

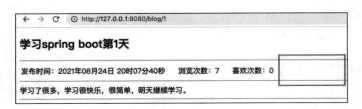

图 11.35　未登录时的博客详情页面

可以看到，未登录状态下没有任何按钮。

回到博客列表，通过登录链接进入登录页面，登录后再次回到第一篇博客，如图11.36所示。

图 11.36　博客详情页面显示喜欢按钮

可以看到出现了"喜欢"按钮。此时喜欢次数为0，单击"喜欢"按钮，喜欢次数显示为1，同时"喜欢"按钮变成"取消喜欢"按钮，如图11.37所示。

图 11.37　博客详情页面显示取消喜欢按钮

通过数据库可视化工具可以查看到表user_liked_blog新创建的记录，如图11.38所示。

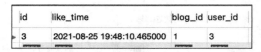

图 11.38　在数据库中查询博客喜欢记录

单击"取消喜欢"按钮，喜欢次数变为0，并且"取消喜欢"按钮变回"喜欢"按钮，如图11.39所示。

图 11.39　"取消喜欢"按钮变回"喜欢"按钮

再次查看数据库,表user_liked_blog中上一条创建的记录也被清除掉了,如图11.40所示。

图 11.40　数据库数据被清除

11.8　配置Spring Security访问规则

前面为了方便调试,将Spring Security的所有接口和页面都设置为允许免登录访问。完成开发之后,可以对接口设置访问规则,并初始化管理员用户。

11.8.1　创建管理员用户

首先使用SQL创建一个管理员用户,并指定password。这里注意password的密文可以使用PasswordEncoder的Encode方法生成,比如可以使用如下代码在控制台输出密码为"abc123"的密文:

```
package com.example.selfblog;

import org.junit.jupiter.api.Test;
import org.springframework.beans.factory.annotation.Autowired;
import org.springframework.boot.test.context.SpringBootTest;
import org.springframework.security.crypto.password.PasswordEncoder;

@SpringBootTest
class SelfblogApplicationTests {
@Autowired
    PasswordEncoder passwordEncoder;
    @Test
    void contextLoads() {
        System.out.println(passwordEncoder.encode("abc123"));
    }
}
```

执行测试方法contextLoads,控制台输出如下密文:

```
$2a$10$ZOTi3oUw9ccgBF3G8cTQhe91Y5RP0Bz7ISbrvqmEwpTDCzwiG0vXu
```

注意,每次执行的代码即使明文一样,密文也不一样。

定义管理员用户手机号为12345678901,并设置密码为abc123,所以密文就是上面的输出,由此可以定义在数据库中插入管理员用户的SQL代码如下:

```
insert into users(password, phone_number, regi_time, update_time)
    values('$2a$10$ZOTi3oUw9ccgBF3G8cTQhe91Y5RP0Bz7ISbrvqmEwpTDCzwiG0vXu',
'12345678901', now(), now());
```

在数据库中执行这条SQL语句,即可创建管理员用户。

11.8.2 配置接口的访问权限

这里应该限制除两个登录接口和发送验证码接口之外其他所有接口的访问权限，可以通过config.security包下的Config配置类来完成。

之前放开访问权限就是通过这个Config配置类的configure(HttpSecurity http)方法来实现的，这里修改此方法，修改后代码如下：

```
@Override
protected void configure(HttpSecurity http) throws Exception {
    http.authorizeRequests().antMatchers("/js/**", "/login.html", "/login/**", "/").permitAll()
        .and().authorizeRequests().anyRequest().authenticated();
    http.csrf().disable();
}
```

上面的代码中，将JS资源、静态页面/login.html、系统根目录和所有以/login开头的接口设置成了允许未登录用户访问，其他接口都必须登录后才能访问。

配置后启动项目，在浏览器中访问http://127.0.0.1:8080/login.html，访问如常，如图11.41所示。

图11.41　设置权限后访问登录页面

再访问项目根目录http://127.0.0.1:8080/，也可以访问，并且从页面上可以看到当前是未登录的状态，如图11.42所示。

图11.42　设置权限后访问列表页面

单击第一篇博客，浏览器没有打开博客详情页面，提示403，如图11.43所示。

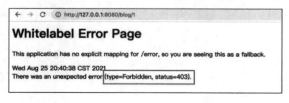

图11.43　无法访问博客详情页

这是因为上面的配置路径中没有配置/blog下免登录访问。

在上面的configure(HttpSecurity http)方法中增加配置，允许"/blog/*"的GET请求访问，增加的代码如下：

```
.and().authorizeRequests().antMatchers(HttpMethod.GET,
"/blog/*").permitAll()
```

增加配置后，configure(HttpSecurity http)方法代码如下：

```
@Override
protected void configure(HttpSecurity http) throws Exception {
    http.authorizeRequests().antMatchers("/js/**", "/login.html",
"/login/**", "/").permitAll()
            .and().authorizeRequests().antMatchers(HttpMethod.GET,
"/blog/*").permitAll()
            .and().authorizeRequests().anyRequest().authenticated();
    http.csrf().disable();
}
```

然后重启项目，在浏览器中访问第一篇博客详情，地址为http://127.0.0.1:8080/blog/1，已经可以访问了，如图11.44所示。

图11.44　设置权限后访问博客详情页

11.8.3　配置仅管理员用户可以发布博客

上面的配置并没有限制发布博客页面和接口仅支持管理员访问，这一节就实现一下限制仅允许管理员发布博客。

实现的方式有很多，这里在代码中使用hasAuthority来限制。将管理员的Authority定义为ADMIN，修改configure(HttpSecurity http)方法的代码，增加两行配置代码，代码如下：

```
.and().authorizeRequests().antMatchers(HttpMethod.GET,
"/publish_blog.html").hasAuthority("ADMIN")
.and().authorizeRequests().antMatchers(HttpMethod.PUT,
"/blog").hasAuthority("ADMIN")
```

增加代码后，configure(HttpSecurity http)方法完整代码如下：

```
@Override
protected void configure(HttpSecurity http) throws Exception {
    http.authorizeRequests().antMatchers("/js/**", "/login.html",
"/login/**", "/").permitAll()
```

```
                .and().authorizeRequests().antMatchers(HttpMethod.GET,
"/blog/*").permitAll()
                .and().authorizeRequests().antMatchers(HttpMethod.GET,
"/publish_blog.html").hasAuthority("ADMIN")
                .and().authorizeRequests().antMatchers(HttpMethod.PUT,
"/blog").hasAuthority("ADMIN")
                .and().authorizeRequests().anyRequest().authenticated();
    http.csrf().disable();
}
```

上面是将HTTP接口限制权限，还需要给管理员12345678901定义权限才可以。

在User类中定义方法getAuthorities()，此方法根据phoneNumber判断，如果是管理员12345678901时增加权限ADMIN，具体代码如下：

```
public Collection<? extends GrantedAuthority> getAuthorities() {
    List<SimpleGrantedAuthority> authorities = new
ArrayList<SimpleGrantedAuthority>(0);
    if("12345678901".equals(this.phoneNumber)){
        authorities.add(new SimpleGrantedAuthority("ADMIN"));
    }
    return authorities;
}
```

然后需要框架能获取到管理员权限，一是在LoginServiceImpl的updateSecurityUserInfo方法中生成UsernamePasswordAuthenticationToken实例时设置Authorities，修改后方法updateSecurityUserInfo代码如下：

```
public String updateSecurityUserInfo(User user) {
    // 更新Security登录用户的信息
    UsernamePasswordAuthenticationToken usernamePasswordAuthenticationToken
            = new UsernamePasswordAuthenticationToken(user, null,
user.getAuthorities());
    SecurityContextHolder.getContext().setAuthentication
(usernamePasswordAuthenticationToken);

    String token = tokenService.generateToken(user);
    return token;
}
```

二是在TokenFilter的doFilterInternal方法中生成UsernamePasswordAuthenticationToken实例时设置Authorities，修改后doFilterInternal方法代码如下：

```
@Override
protected void doFilterInternal(HttpServletRequest request,
HttpServletResponse response, FilterChain filterChain) throws ServletException,
IOException {
    String token = request.getHeader("token");
    if(token!=null && !"".equals(token.trim())){//
&& !token.startsWith("JSESSIONID")){
        Integer userId = tokenService.getUserIdFromToken(token);
        // 可以从Token中拿到用户名，但是未登录
```

```
            if (null != userId && null == SecurityContextHolder.getContext().
getAuthentication()) {
                User user = null;
                try {
                    user = userRepository.getById(userId);
                } catch (JpaObjectRetrievalFailureException e) {
                    log.error("token 中的 userId 无效" + userId, e);
                    throw new RuntimeException("token 中的 userId 无效" + userId);
                }
                if (user != null && tokenService.validateToken(token, user)) {
                    UsernamePasswordAuthenticationToken usernamePasswordAuthenticationToken = new UsernamePasswordAuthenticationToken(user, null, user.getAuthorities());
                    usernamePasswordAuthenticationToken.setDetails(new WebAuthenticationDetailsSource().buildDetails(request));
                    SecurityContextHolder.getContext().setAuthentication(usernamePasswordAuthenticationToken);
                } else {
                    log.error("登录状态失效，请重新登录。");
                    throw new RuntimeException("登录状态失效，请重新登录。");
                }
            }
        }
        filterChain.doFilter(request, response);
    }
```

11.8.4 测试发布博客权限管理

使用管理员账号1234567890和密码abc123登录，登录成功后在浏览器地址栏访问地址http://127.0.0.1:8080/publish_blog.html，成功打开发布博客页面，如图11.45所示。

在页面中录入博客标题和内容，如图11.46所示。

图 11.45　配置权限后访问发布博客页面

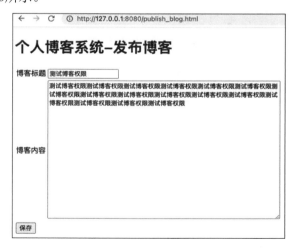

图 11.46　录入博客标题和内容

单击"保存"按钮,浏览器提示"发布博客成功"。然后回到博客列表页面,单击"下一页",可以看到刚发布的博客,如图11.47所示。

图11.47　在博客列表页查看发布的博客

到这里说明管理员拥有发布博客的权限,但是不能说明普通用户不能发布博客。接下来测试普通用户发布博客。

重启项目,访问登录页面地址http://127.0.0.1:8080/login.html,然后随便使用一个手机号登录,只要不是12345678901即可。提示登录成功后,在浏览器访问发布博客的地址http://127.0.0.1:8080/publish_blog.html,不能打开发布博客页面,浏览器提示403,如图11.48所示。

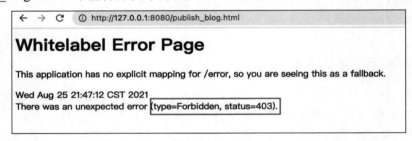

图11.48　发布博客页面不允许普通用户访问

如此说明权限配置准确无误。

第 12 章

Spring Boot项目的测试和部署

项目在开发完成后，通常可能会部署在本机单元测试，部署到测试服务器以供测试人员进行测试，以及最终部署到生产环境正式提供服务。本章将介绍Spring Boot项目在测试和部署中可能使用到的技术。

本章主要涉及的知识点有：

- 如何指定外部配置文件。
- 掌握Spring Profiles的使用。
- 如何使用Spring Boot的测试功能。
- 如何打包和部署Spring Boot应用程序。

12.1 配置的切换

一套项目代码可能会有很多配置项，同一个配置项在不同环境下可能会有不同的配置值。尤其是在项目比较复杂，配置项众多的情况下，纯手工来维护这些配置项会是一件比较烦琐的工作，而且容易出错。而Spring Profiles提供了通过配置profile来分类别、分环境维护和管理系统配置项的功能。本节将介绍Spring Profiles的使用。

在实际项目中，尤其是企业级项目，一套代码至少会部署在开发、测试和生产3套环境中，不同环境下的配置通常不会完全相同。如此一来，需要在不同环境下指定不同的配置。

在不同环境下指定不同配置参数的方法有很多，甚至Profiles功能也不是Spring独有的，Maven也有它的Profile功能。在介绍Spring Profiles之前，先来介绍一种使用外部配置文件的方法。

12.1.1 在项目启动时指定外部配置文件

Spring Boot项目启动时，会默认自动加载类路径目录下的application.yaml和application.properties文件作为配置文件，这个配置文件的路径可以通过配置项spring.config.location来指定。

配置项spring.config.location可以在项目启动时通过命令参数"-D"来指定。比如启动/root/demo1.jar，指定配置文件为系统文件/root/demo1.yaml，那么启动时命令行命令如下：

```
java -Dspring.config.location=/root/demo.yaml -jar /root/demo1.jar
```

【示例12.1】 通过命令行启动项目，指定外部配置文件

为了演示这个配置项的使用，先来创建一个空项目，项目命名为demo1。项目demo为Spring Boot项目，除了spring-boot-starter和spring-boot-starter-test外，只依赖lombok，pom.xml文件代码如下：

```xml
<?xml version="1.0" encoding="UTF-8"?>
<project xmlns="http://maven.apache.org/POM/4.0.0" xmlns:xsi="http://www.w3.org/2001/XMLSchema-instance"
    xsi:schemaLocation="http://maven.apache.org/POM/4.0.0 https://maven.apache.org/xsd/maven-4.0.0.xsd">
    <modelVersion>4.0.0</modelVersion>
    <parent>
        <groupId>org.springframework.boot</groupId>
        <artifactId>spring-boot-starter-parent</artifactId>
        <version>2.5.4</version>
        <relativePath/> <!-- lookup parent from repository -->
    </parent>
    <groupId>com.example</groupId>
    <artifactId>demo1</artifactId>
    <version>0.0.1-SNAPSHOT</version>
    <name>demo1</name>
    <description>Demo project for Spring Boot</description>
    <properties>
        <java.version>11</java.version>
    </properties>
    <dependencies>
        <dependency>
            <groupId>org.springframework.boot</groupId>
            <artifactId>spring-boot-starter</artifactId>
        </dependency>

        <dependency>
            <groupId>org.projectlombok</groupId>
            <artifactId>lombok</artifactId>
            <optional>true</optional>
        </dependency>
        <dependency>
            <groupId>org.springframework.boot</groupId>
            <artifactId>spring-boot-starter-test</artifactId>
            <scope>test</scope>
        </dependency>
    </dependencies>

    <build>
        <plugins>
            <plugin>
```

```xml
            <groupId>org.springframework.boot</groupId>
            <artifactId>spring-boot-maven-plugin</artifactId>
            <configuration>
                <excludes>
                    <exclude>
                        <groupId>org.projectlombok</groupId>
                        <artifactId>lombok</artifactId>
                    </exclude>
                </excludes>
            </configuration>
        </plugin>
      </plugins>
    </build>
</project>
```

创建一个配置文件类，用于读取配置文件。创建config目录，并在config目录下创建ProfileConfigProperties，具体代码如下：

```
package com.example.demo1.config;

import lombok.Data;
import org.springframework.boot.context.properties.ConfigurationProperties;
import org.springframework.stereotype.Component;

@Component
@Data
@ConfigurationProperties("test.user")
public class ProfileConfigProperties {
    private String name;
    private Integer age;
}
```

为了能自动被容器加载，使用了@Component注解。配置类ProfileConfigProperties会在容器初始化时被实例化，并且实例的name和age字段分别会被赋予配置项test.user.name和配置项test.user.age的值。

为了看到实例的字段信息，对程序启动类Demo1Application的main方法进行改造，获取并打印ProfileConfigProperties的实例。改造后，程序启动类Demo1Application的代码如下：

```
package com.example.demo1;

import com.example.demo1.config.ProfileConfigProperties;
import org.springframework.boot.SpringApplication;
import org.springframework.boot.autoconfigure.SpringBootApplication;
import org.springframework.context.ConfigurableApplicationContext;

@SpringBootApplication
public class Demo1Application {
    public static void main(String[] args) {
        ConfigurableApplicationContext applicationContext = SpringApplication.run(Demo1Application.class, args);
```

```
            ProfileConfigProperties bean = applicationContext.getBean
(ProfileConfigProperties.class);
            System.out.println(bean);
        }
    }
```

为了在命令行中启动程序，先给程序打包。通过maven命令来打包项目，首先打开IDEA的Terminal面板，如图12.1所示。

打开后默认会在项目目录下，如图12.2所示。

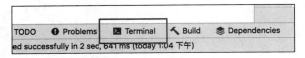

图12.1　IDEA 的 Terminal 面板

图12.2　命令行中的默认目录

输出Maven的打包命令，命令代码如下：

```
mvn package
```

按回车键执行命令，注意这里需要安装Maven。等待Maven对代码进行编译、测试和打包，若输出如图12.3所示，则表示打包成功。

```
[INFO] Tests run: 1, Failures: 0, Errors: 0, Skipped: 0
[INFO]
[INFO]
[INFO] --- maven-jar-plugin:3.2.0:jar (default-jar) @ demo1 ---
[INFO]
[INFO] --- spring-boot-maven-plugin:2.5.4:repackage (repackage) @ demo1 ---
[INFO] Replacing main artifact with repackaged archive
[INFO] ------------------------------------------------------------------------
[INFO] BUILD SUCCESS
[INFO] ------------------------------------------------------------------------
[INFO] Total time:  6.937 s
[INFO] Finished at: 2021-08-26T16:02:11+08:00
[INFO] ------------------------------------------------------------------------
```

图12.3　打包成功的日志输出

打开Project面板中的target目录，如图12.4所示，可以看到打包好的可执行JAR包。

回到Terminal面板，执行命令"cd target"进入target目录。先不指定外部配置文件，使用如下命令执行生成的JAR包：

```
java -jar demo1-0.0.1-SNAPSHOT.jar
```

可以看到输出如图12.5所示，成功打印了ProfileConfigProperties实例，只是因为没有配置文件指定配置项test.user.name和配置项test.user.age，所以实例属性值为空。

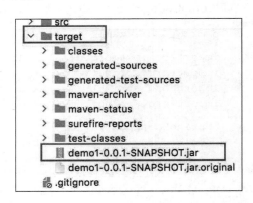

图12.4　程序打包后的 target 目录

```
ProfileConfigProperties(name=null, age=null)
```

图 12.5　打印 ProfileConfigProperties 实例

然后在系统目录/root下创建名为demo1.yaml的文件，并在文件中配置如下：

```
test:
  user:
    name: zhangsan
    age: 18
```

然后指定再次运行JAR包，并指定demo1.yaml作为配置文件。在Terminal中执行如下命令代码：

```
java -Dspring.config.location=/root/demo1.yaml -jar demo1-0.0.1-SNAPSHOT.jar
```

程序执行结束，输出如图12.6所示，可以看到实例被注入了外部配置文件的配置值。

```
ProfileConfigProperties(name=zhangsan, age=18)
```

图 12.6　打印配置参数

注　意
　　-Dspring.config.location在-jar之前，如果放到-jar之后，则指定无效。

对于已经稳定运行的项目，独立的配置文件还是比较方便的，尤其是对于只需要修改配置文件的情况，修改配置文件后只需要重启项目即可，而不用重新编译和打包。

12.1.2　Spring Profile 的使用

使用Spring Profile的大致思路是将配置分组，每一组和一个Profile名绑定，然后在项目配置文件中或者启动时指定某一个或多个Profile名，从而相应地激活与Profile绑定的配置项目。

常用的Profile名有dev、test和prod，分别表示开发环境、测试环境和生产环境。

激活当前的Profile通过配置项spring.profiles.active来实现，比如通过如下代码来激活开发环境dev：

```
spring:
  profiles:
    active: dev
```

配置项spring.profiles.active支持使用英文逗号分隔来同时激活多个Profile环境。比如同时激活dev和mysql，代码如下：

```
spring:
  profiles:
    active: dev,mysql   #这里的 mysql 是 profile 名
```

1. 文件中配置项的 profile 划分

Spring Boot默认除了会加载application.properties或application.yaml之外，还会加载Profile

特定的配置文件。Profile特定配置文件的命名规则是application-{profile}，即如果配置项spring.profiles.active的值是dev，那么默认会加载application-dev.properties或application-dev.yaml。如果指定激活多个Profile，则会在加载application.properties或application.yaml之外加载多个application-{profile}配置文件。

同时加载多个配置文件就有可能出现对一个配置项多次指定配置值的情况，此时会有一个配置值覆盖顺序。在application.properties或application.yaml中的配置会被application-{profile}中的覆盖。如果多个application-{profile}中定义了相同的配置，那么会根据配置项spring.profiles.active所指定的顺序，后面的覆盖前面的。

【示例12.2】 验证配置的覆盖顺序

仍然使用上一个示例的项目，在resources目录下创建两个配置文件application-dev.yaml和application-mysql.yaml。在application-dev.yaml文件中定义name为lisi，age为19，配置如下：

```
test:
  user:
    name: lisi
    age: 19
```

在application-mysql.yaml中将name设置为wangwu，age设置为20，配置如下：

```
test:
  user:
    name: wangwu
    age: 20
```

而application.yaml中仍然是原先的zhangsan和18。现在通过配置项spring.profiles.active来同时激活dev和mysql，配置如下：

```
spring:
  profiles:
    active: dev,mysql
test:
  user:
    name: zhangsan
    age: 18
```

运行项目，控制台输出的是wangwu和20，如图12.7所示。

现在切换配置项spring.profiles.active的顺序，指定如下：

```
spring:
  profiles:
    active: mysql,dev
```

运行项目，程序输出为lisi和19，如图12.8所示。

```
ProfileConfigProperties(name=wangwu, age=20)
```

```
ProfileConfigProperties(name=lisi, age=19)
```

图 12.7 不同 Profile 的配置覆盖结果　　　图 12.8 切换 Profile 顺序后的配置项覆盖结果

上面是通过配置文件将配置项分组，还可以在一个文件内定义不同的Profile的属性。在application.yaml中可以通过3个短横线"---"来将配置项分组，在每一组内使用配置项spring.config.activate.on-profile来指定当前组配置所属的Profile。这种配置的好处是简便，无须创建多个配置文件，但缺点是不具有根据spring.profiles.active的配置顺序来进行配置项覆盖的功能。在同一个文件中，配置项的覆盖规则是后面的配置覆盖前面的配置。

2. 配置类中使用注解@Profile

在代码中定义的组件、配置类和配置属性（@Component、@Configuration或@ConfigurationProperties），还可以使用注解@Profile来标注这些配置代码，使这些代码只在指定的Profile下才生效。

注解@Profile的value是String数组类型，所以可以指定一个或多个Profile名，使得被标注的组件或配置在被指定的多个Profile中的任何一个激活时才生效。也就是说，注解@Profile指定多个Profile时，它们是"或"的关系。

注解@Profile和@Component、@Configuration使用时一起标注在同一个类或方法上。不过，对于注解@ConfigurationProperties定义的配置属性类，如果它是通过注解@EnableConfigurationProperties来注册的，那么应该将注解@Profile标注到@EnableConfigurationProperties所标注的类上才生效。

12.2 Spring Boot的测试功能

使用Spring Initializr生成的项目会默认添加spring-boot-starter-test依赖，并且会在src/test/java目录下自动生成一个测试类，在生成的测试类中可以完成许多单元测试工作。本节将介绍spring-boot-starter-test支持的测试功能。

12.2.1 构建测试类

使用Spring Initializr自动生成的测试类都会被@SpringBootTest标注，在类中定义方法contextLoads，方法被junit的@Test注解标注。被标注的类拥有单元测试的功能，其中使用@Test注解标注的方法可以在IDE中通过右键快捷菜单Run来运行。

【示例12.3】 运行demo1项目的单元测试contextLoads方法

项目demo1是示例12.1中通过Spring Initializr创建的项目，项目被创建时自动生成了测试类Demo1ApplicationTests。测试类Demo1ApplicationTests在项目中的目录结构如图12.9所示。

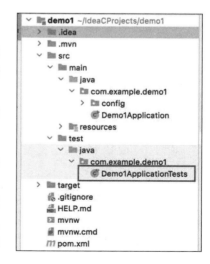

图12.9　项目目录结构

测试类Demo1ApplicationTests默认会包含方法contextLoads，并且方法被@Test注解标注。测试类Demo1ApplicationTests具体代码如下：

```java
package com.example.demo1;

import org.junit.jupiter.api.Test;
import org.springframework.boot.test.context.SpringBootTest;

@SpringBootTest
class Demo1ApplicationTests {

    @Test
    void contextLoads() {
    }

}
```

在方法contextLoads上右击，选择Run 'contextLoads()'，运行程序。虽然方法contextLoads的方法体没有代码，但是仍然可以在控制台看到大量的日志输出，甚至是Spring Boot应用启动成功的Logo，日志输出如图12.10所示。

```
10:19:25.571 [main] DEBUG org.springframework.test.context.support.TestPropertySourceUtils -
Adding inlined properties to environment: {spring.jmx.enabled=false, org.springframework.boot
.test.context.SpringBootTestContextBootstrapper=true}

  .   ____          _            __ _ _
 /\\ / ___'_ __ _ _(_)_ __  __ _ \ \ \ \
( ( )\___ | '_ | '_| | '_ \/ _` | \ \ \ \
 \\/  ___)| |_)| | | | | || (_| |  ) ) ) )
  '  |____| .__|_| |_|_| |_\__, | / / / /
 =========|_|==============|___/=/_/_/_/
 :: Spring Boot ::        (v2.5.4)

2021-08-27 10:19:26.286  INFO 931 --- [           main] com.example.demo1.Demo1ApplicationTests
  : Starting Demo1ApplicationTests using Java 11.0.11 on mikedeair with PID 931 (started by
mike in /Users/mike/IdeaCProjects/demo1)
2021-08-27 10:19:26.289  INFO 931 --- [           main] com.example.demo1.Demo1ApplicationTests
  : The following profiles are active: dev
2021-08-27 10:19:27.126  INFO 931 --- [           main] com.example.demo1.Demo1ApplicationTests
  : Started Demo1ApplicationTests in 1.547 seconds (JVM running for 3.198)
```

图 12.10　测试方法运行的日志输出

如果在Run面板中看不到日志输出，可以通过控制台左侧的show passed来查看所属不同部分的日志，如图12.11所示，选中Test Results看到的就是完整的日志输出。

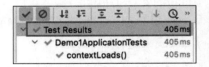

图 12.11　日志分类面板

从日志输出中也可以看到，这里单元测试时相当于启动了完整的Spring容器。启动Spring容器的功能是借助注解@SpringBootTest来实现的，如果是Web项目，还可以通过配置注解@SpringBootTest的属性来启动Web容器。下面详细介绍注解@SpringBootTest的使用。

注解@SpringBootTest提供了4个属性来对测试环境进行自定义，这4个属性是properties、args、classes和webEnvironment。

（1）属性properties也是默认的values属性，String数组类型，用key=value的格式来指定Spring Environment的属性值。

【示例12.4】 使用注解@SpringBootTest的properties指定ProfileConfigProperties的name值在测试类Demo1ApplicationTests中注入ProfileConfigProperties，代码如下：

```
@Autowired
ProfileConfigProperties profileConfigProperties;
```

在测试方法contextLoads中打印注入的实例，代码如下：

```
@Test
void contextLoads() {
    System.out.println(profileConfigProperties);
}
```

运行测试方法contextLoads，控制台方法输出如图12.12所示。

```
ProfileConfigProperties(name=lisi, age=19)
```

图 12.12　测试的配置项打印

这是配置文件中配置的值，然后通过@SpringBootTest注解的properties属性将test.user.name指定为test，代码如下：

```
@SpringBootTest(properties = "test.user.name=test")
```

再次运行测试方法contextLoads，这次输出结果age仍然为配置文件中配置的19，但是name已经成为通过属性properties指定的test，控制台日志如图12.13所示。

```
ProfileConfigProperties(name=test, age=19)
```

图 12.13　通过 properties 指定配置项

由此可以看出，通过properties属性设置的配置项会覆盖配置文件中的配置。

（2）第二个属性args用来指定程序的启动参数。通过这个属性设置的参数会传递给SpringApplication的run方法。

（3）第三个属性classes是Class数组类型，指定用来加载ApplicationContext的组件类，也可以通过注解@ContextConfiguration的classes属性来指定。

（4）最后一个属性webEnvironment是枚举WebEnvironment类型，指定运行测试方法时启动一个什么样的Web环境，默认值是WebEnvironment.MOCK类型。

当属性webEnvironment为WebEnvironment.MOCK类型时又分为几种情况，如果类路径下有Servlet APIs，那么将会创建一个使用伪（mock）Servlet环境的WebApplicationContext；如果类路径下有Spring WebFlux，那么将会创建ReactiveWebApplicationContext；其他情况将会创建ApplicationContext。上面的示例属于最后一种情况，创建的是ApplicationContext。

RANDOM_PORT是枚举类WebEnvironment的另一个值，指定为此值会创建一个Web应用程序上下文，从而提供一个真实的Web环境，并且会通过设置server.port=0来触发对一个随机端口的监听。使用Spring Initializr创建项目时，如果依赖了Spring Web，那么创建的测试类中注解@ContextConfiguration的属性webEnvironment会被指定为RANDOM_PORT。

枚举类WebEnvironment还有另外两个枚举值：DEFINED_PORT和NONE。DEFINED_PORT和RANDOM_PORT类似，只是DEFINED_PORT会使用配置文件中指定的服务端口，而NONE不会创建Web环境。

如果项目不是Web项目，将webEnvironment指定为DEFINED_PORT或者RANDOM_PORT，那么在启动测试类时会报错。例如demo1中测试类指定为RANDOM_PORT，代码如下：

```
@SpringBootTest(webEnvironment = SpringBootTest.WebEnvironment.RANDOM_PORT)
```

然后运行测试方法contextLoads，控制台报错如图12.14所示。

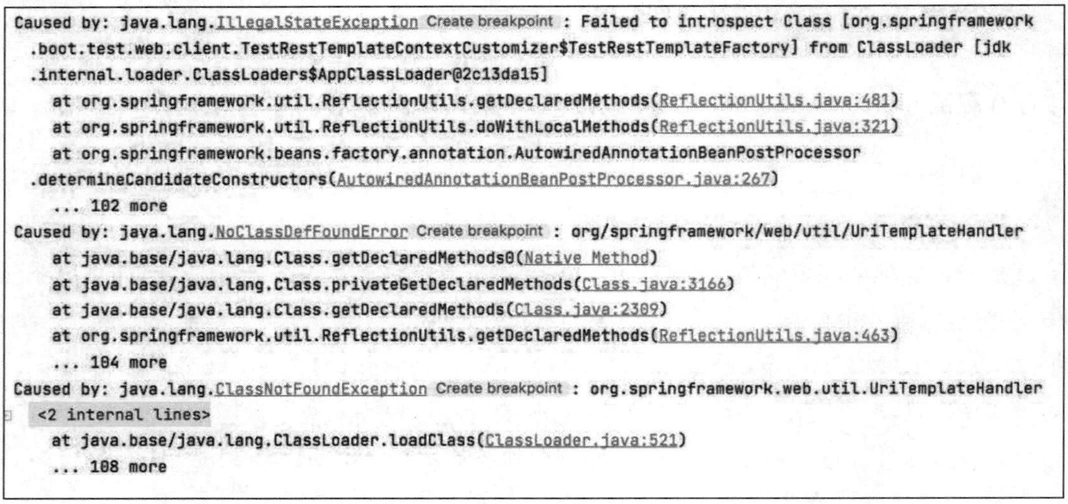

```
Caused by: java.lang.IllegalStateException Create breakpoint : Failed to introspect Class [org.springframework
.boot.test.web.client.TestRestTemplateContextCustomizer$TestRestTemplateFactory] from ClassLoader [jdk
.internal.loader.ClassLoaders$AppClassLoader@2c13da15]
    at org.springframework.util.ReflectionUtils.getDeclaredMethods(ReflectionUtils.java:481)
    at org.springframework.util.ReflectionUtils.doWithLocalMethods(ReflectionUtils.java:321)
    at org.springframework.beans.factory.annotation.AutowiredAnnotationBeanPostProcessor
.determineCandidateConstructors(AutowiredAnnotationBeanPostProcessor.java:267)
    ... 102 more
Caused by: java.lang.NoClassDefFoundError Create breakpoint : org/springframework/web/util/UriTemplateHandler
    at java.base/java.lang.Class.getDeclaredMethods0(Native Method)
    at java.base/java.lang.Class.privateGetDeclaredMethods(Class.java:3166)
    at java.base/java.lang.Class.getDeclaredMethods(Class.java:2309)
    at org.springframework.util.ReflectionUtils.getDeclaredMethods(ReflectionUtils.java:463)
    ... 104 more
Caused by: java.lang.ClassNotFoundException Create breakpoint : org.springframework.web.util.UriTemplateHandler
    <2 internal lines>
    at java.base/java.lang.ClassLoader.loadClass(ClassLoader.java:521)
    ... 108 more
```

图 12.14　错误指定 webEnvironment 时的报错信息

不妨在pom.xml中将依赖spring-boot-starter修改成spring-boot-starter-web：

```
<dependency>
    <groupId>org.springframework.boot</groupId>
    <artifactId>spring-boot-starter-web</artifactId>
</dependency>
```

刷新Maven依赖，然后运行测试方法contextLoads，控制台输出如图12.15所示。

可以看到，Tomcat初始化时端口为0，最后以端口61071启动。再次运行测试方法contextLoads，控制台输出如图12.16所示。

```
2021-08-27 15:56:14.383  INFO 1909 --- [           main] o.s.b.w.embedded.tomcat.TomcatWebServer  : Tomcat
initialized with port(s): 0 (http)
2021-08-27 15:56:14.398  INFO 1909 --- [           main] o.apache.catalina.core.StandardService   : Starting
service [Tomcat]
2021-08-27 15:56:14.398  INFO 1909 --- [           main] org.apache.catalina.core.StandardEngine  : Starting
Servlet engine: [Apache Tomcat/9.0.52]
2021-08-27 15:56:14.504  INFO 1909 --- [           main] o.a.c.c.C.[Tomcat].[localhost].[/]       :
Initializing Spring embedded WebApplicationContext
2021-08-27 15:56:14.504  INFO 1909 --- [           main] w.s.c.ServletWebServerApplicationContext : Root
WebApplicationContext: initialization completed in 1891 ms
2021-08-27 15:56:15.721  INFO 1909 --- [           main] o.s.b.w.embedded.tomcat.TomcatWebServer  : Tomcat
started on port(s): 61071 (http) with context path ''
2021-08-27 15:56:15.739  INFO 1909 --- [           main] com.example.demo1.Demo1ApplicationTests  : Started
Demo1ApplicationTests in 3.928 seconds (JVM running for 5.521)
```

图 12.15　初始化端口和随机监听的端口

```
WebApplicationContext: initialization completed in 1766 ms
2021-08-27 16:00:59.117  INFO 1936 --- [           main] o.s.b.w.embedded.tomcat.TomcatWebServer  : Tomcat
started on port(s): 61096 (http) with context path ''
2021-08-27 16:00:59.129  INFO 1936 --- [           main] com.example.demo1.Demo1ApplicationTests  : Started
```

图 12.16　随机监听的端口

Tomcat启动的端口变成了61096。

12.2.2　测试的自动配置

Spring Boot的自动配置功能非常强大，但是使用单元测试，尤其是在开发中使用单元测试通常可能只是测试某一部分功能，这个时候如果每次执行一个测试方法，Spring Boot都要进行全量的自动配置，对于略微复杂一些的项目来说，都是比较耗时的。为了一个简单的输出，要等上十多秒甚至几十秒的启动时间，在开发中频繁使用单元测试是非常低效率的。

spring-boot-test提供了测试一个slice的功能，就是在测试时只配置应用程序的一部分。使用slice测试功能的好处是除了启动速度更快外，对于测试环境的依赖也降低了。比如，在引入了JPA后，容器初始化会自动配置数据库连接，如果数据库无法连接，就会导致容器初始化失败。如此一来，对于全量配置，即使只测试一个简单的功能，也要全部的服务配合，但这样的环境在开发机器上不一定能够支持。

这一功能可以通过使用一系列以Test为后缀的注解来使用，如@DataJdbcTest、@WebMvcTest、@JsonTest等。

1. spring-boot-test 提供的 slice 注解

这些注解是spring-bot-test提供的，在spring-boot-test-autoconfigure包下，总共有15个注解，分别为@DataCassandraTest、@DataJdbcTest、@DataJpaTest、@DataLdapTest、@DataMongoTest、@DataNeo4jTest、@DataR2dbcTest、@DataRedisTest、@JdbcTest、@JooqTest、@JsonTest、@RestClientTest、@WebFluxTest、@WebMvcTest和@WebServiceClientTest。

从注解的命名大致可以看出这些注解所属的slice，前面的10个，即@DataCassandraTest、@DataJdbcTest、@DataJpaTest、@DataLdapTest、@DataMongoTest、@DataNeo4jTest、@DataR2dbcTest、@DataRedisTest、@JdbcTest和@JooqTest都是和数据库相关的技术。后面的

@JsonTest是对JSON的自动配置，这个类的自动配置列举在表12.1中。@RestClientTest会自动配置REST Clients。@WebFluxTest、@WebMvcTest是对Web的自动配置，@WebMvcTest会导入的自动配置类列举在了表12.1中。最后的@WebServiceClientTest是对使用了Spring Web Services技术项目的自动配置。

表 12.1 部分测试 slice 注解的导入说明

注 解 名	导入自动生效配置（以下的基础包为 org.springframework.boot.autoconfigure）
@JsonTest	cache.CacheAutoConfiguration gson.GsonAutoConfiguration jackson.JacksonAutoConfiguration jsonb.JsonbAutoConfiguration org.springframework.boot.test.autoconfigure.json.JsonTestersAutoConfiguration
@WebMvcTest	cache.CacheAutoConfiguration context.MessageSourceAutoConfiguration data.web.SpringDataWebAutoConfiguration freemarker.FreeMarkerAutoConfiguration groovy.template.GroovyTemplateAutoConfiguration gson.GsonAutoConfiguration hateoas.HypermediaAutoConfiguration http.HttpMessageConvertersAutoConfiguration jackson.JacksonAutoConfiguration jsonb.JsonbAutoConfiguration mustache.MustacheAutoConfiguration security.oauth2.client.servlet.OAuth2ClientAutoConfiguration
@WebMvcTest	security.oauth2.resource.servlet.OAuth2ResourceServerAutoConfiguration security.servlet.SecurityAutoConfiguration security.servlet.SecurityFilterAutoConfiguration security.servlet.UserDetailsServiceAutoConfiguration task.TaskExecutionAutoConfiguration thymeleaf.ThymeleafAutoConfiguration validation.ValidationAutoConfiguration web.servlet.HttpEncodingAutoConfiguration web.servlet.WebMvcAutoConfiguration web.servlet.error.ErrorMvcAutoConfiguration org.springframework.boot.test.autoconfigure.web.servlet.MockMvcAutoConfiguration org.springframework.boot.test.autoconfigure.web.servlet.MockMvcSecurityConfiguration org.springframework.boot.test.autoconfigure.web.servlet.MockMvcWebClientAutoConfiguration org.springframework.boot.test.autoconfigure.web.servlet.MockMvcWebDriverAutoConfiguration

Spring Boot官方文档中列举了所有注解自动配置的内容，具体是在官方文档Spring Boot Reference Documentation附录的Test Auto-configuration Classes部分，如图12.17所示。

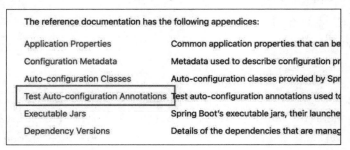

图 12.17 官方文档的测试自动配置注解列表

2．slice 相关注解的使用

上面列举了spring-boot-test提供的slice测试注解，接下来介绍这些注解的使用。

这些注解需要标注在类上，和注解@SpringBootTest是一个级别的，被标注的类成为测试类。然后创建测试方法，使用测试方法和普通单元测试方法一样。下面以@JsonTest为例进行说明。

【示例12.5】 使用JsonTest测试JSON日期格式配置

在测试目录下的com.example.demo1中创建测试类JsonTests，并使用@JsonTest对类标注，代码如下：

```
package com.example.demo1;

import org.springframework.boot.test.autoconfigure.json.JsonTest;

@JsonTest
public class JsonTests {}
```

在类中创建测试方法test，无参，并使用注解@Test标注。为了证明JsonTests仅会自动配置JSON相关的配置，不妨先在类中注入前面示例中用到过的ProfileConfigProperties，在测试类添加注入的代码如下：

```
@Autowired
ProfileConfigProperties profileConfigProperties;
```

然后运行测试方法test。不出所料，测试无法通过，并且控制台报错，如图12.18所示。

```
Caused by: org.springframework.beans.factory.NoSuchBeanDefinitionException Create breakpoint : No qualifying bean of type
  'com.example.demo1.config.ProfileConfigProperties' available: expected at least 1 bean which qualifies as autowire
  candidate. Dependency annotations: {@org.springframework.beans.factory.annotation.Autowired(required=true)}
    at org.springframework.beans.factory.support.DefaultListableBeanFactory.raiseNoMatchingBeanFound
  (DefaultListableBeanFactory.java:1790)
    at org.springframework.beans.factory.support.DefaultListableBeanFactory.doResolveDependency
  (DefaultListableBeanFactory.java:1346)
    at org.springframework.beans.factory.support.DefaultListableBeanFactory.resolveDependency
  (DefaultListableBeanFactory.java:1300)
    at org.springframework.beans.factory.annotation.AutowiredAnnotationBeanPostProcessor$AutowiredFieldElement
  .resolveFieldValue(AutowiredAnnotationBeanPostProcessor.java:657)
    ... 72 more
```

图 12.18　没有初始化 ProfileConfigProperties 时报的错误

这说明创建容器中没有配置类ProfileConfigProperties实例。

删除注入ProfileConfigProperties的两行代码，在测试类中注入JacksonTester，并且在测试方法中使用JacksonTester，将Date对象转为JSON字符串并打印，具体代码如下：

```
@Autowired
private JacksonTester json;
@Test
void test() throws Exception{
   System.out.println(json.write(new Date()).getJson());
}
```

执行测试方法test，测试通过，控制台上输出了默认日期的字符串格式，如图12.19所示。

上面是Jackson对Date类型转换的默认格式，现在来设置转换格式。使用spring.jackson.date-format配置项，将格式设置为只转换为年月日的格式，在application.yaml文件中增加配置代码如下：

```yaml
spring:
  jackson:
    date-format: "yyyy-MM-dd"
```

修改配置后，再次运行测试方法test，控制台的输出结果如图12.20所示。

图12.19　jackson的默认日期格式　　　　图12.20　配置生效后的日志格式

这说明配置生效了，并且使用@JsonTest注解只自动配置了JSON相关的配置。

12.3　打包和部署

打包和部署是解决如何将程序发布到服务器上这一问题的两个步骤，部署方式决定以哪种方式打包或者打包成哪种类型的文件。

12.3.1　打包（JAR和WAR）

JAR和WAR是两种可选的打包文件类型，其中JAR也是Spring Boot默认支持的类型。在使用Spring Initializr创建项目时，Packaging那一项就对应打包类型，默认是JAR，可选WAR，如图12.21所示。

打包类型也可以在生成项目后通过pom.xml文件进行修改。由于JAR也是Maven的默认打包方式，因此在生成JAR的项目时，在pom.xml文件中没有<packaging>标签。

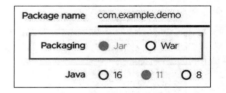

图12.21　Spring Initializr的默认packaging

如果需要将项目打包方式修改成WAR，可以在pom.xml中添加与<groupId>和<artifactId>同级别的packaging标签，配置代码如下：

```
<packaging>war</packaging>
```

如果是使用Spring Initializr生成packaging为WAR的项目，那么即使没有指定添加Web依赖，在生成的pom.xml中也会生成对spring-boot-starter-web的依赖，以及对spring-boot-starter-tomcat的依赖。不过对spring-boot-starter-tomcat依赖的scope为provided，因为WAR包需要部署在Web服务器上，默认Web服务器是Tomcat，而spring-boot-starter-tomcat提供的依赖在部署后由Web服务器提供。

通过Spring Initializr生成的pom.xml文件中都会配置spring-boot-maven-plugin。Spring Boot的Maven插件对于打包有着重要的作用，这是因为Spring Boot项目打包有着其特定的目录结构。

【示例12.6】　将demo1项目打包，并查看JAR的内部接口

当前demo1项目代码有程序启动类Demo1Application和配置类ProfileConfigProperties，以及配置文件application.yaml。IDE中的项目结构如图12.22所示。pom.xml的依赖如图12.23所示。

使用Maven的mvn命令进行打包，这里使用参数-DskipTests来跳过测试，具体命令如下：

```
mvn package -DskipTests
```

打包成功后，在项目的target目录下生成了JAR文件demo1-0.0.1-SNAPSHOT.jar，对JAR文件进行解压，得到如图12.24所示的目录结构。

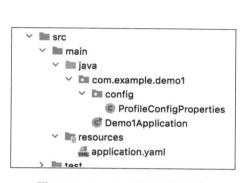

图 12.22　demo1 项目的目录结构

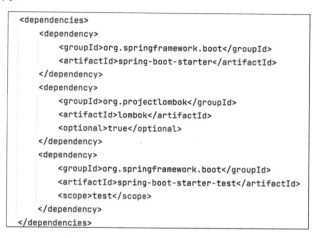

图 12.23　pom.xml 的依赖配置

jar中目录总体上分成3部分，分别说明如下。

（1）第1部分是MANIFEST目录，其中主要是文件MANIFEST.MF，这个文件中的具体内容如图12.25所示。

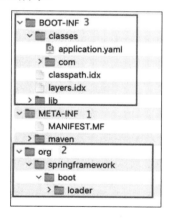

图 12.24　JAR 包的目录结构

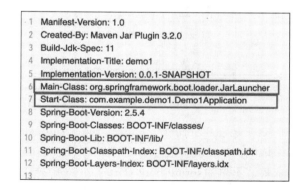

图 12.25　MANIFEST.MF 文件的内容

可以看到其中记录了应用程序的元信息，比如项目名为demo1，还有版本号、JDK版本、Spring Boot版本等。其中最重要的是第6行和第7行，第6行中定义了Main-Class，定义了程序入口。这里的Main-Class为org.springframework.boot.loader.JarLauncher，指向jar目录结构的第2部分。

（2）jar目录接口第2部分的文件是spring-boot-load项目提供的。由于Spring Boot为了打包和部署的简单化，将项目依赖的JAR包都打包到了项目JAR包中，这样也就出现了嵌套JAR包的情况。但是，由于Java官方没有提供嵌套JAR包的规范，因此spring-boot-load项目提供了加载嵌套JAR包的支持。spring-boot-load既会加载项目依赖的JAR包，又会根据MANIFEST.MF中定义的Start-Class来执行应用代码。

（3）在jar目录的第3部分，也就是目录BOOT-INF下，是我们的应用程序代码的class文件和所依赖的JAR包。在class目录下是应用程序的class文件和配置文件，在lib目录下是项目所依赖的JAR包。

【示例12.7】 查看WAR包的目录结构

通过Spring Initializr创建一个packaging为WAR的项目，项目名为demowar，不添加任何依赖。下载到本地后解压，然后用IDEA加载，可以看到项目目录结构如图12.26所示。

程序启动类为DemowarApplication。在DemowarApplication类的main方法中增加一行输出，代码如下：

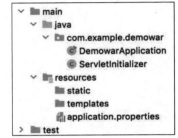

```
System.out.println("demowar 启动成功！");
```

图12.26 demowar 项目目录结构

使用mvn package将项目打包，打包成功后，在target目录下生成文件demowar-0.0.1-SNAPSHOT.war。将文件demowar-0.0.1-SNAPSHOT.war解压，目录如图12.27所示。

从目录中看，同样分成3部分，与JAR包目录只是第3部分不同。JAR包目录下的第3部分目录名是BOOT-INF，这里第3部分目录名是WEB-INF。内部的文件结构基本一致，只是多了一个lib-provided目录，这个目录中存放由Web服务器提供的JAR包。lib-provided目录的出现是为了保证既能在使用jar命令启动WAR包时不缺少依赖，又可以避免项目在Web服务器部署时发生依赖冲突。打开lib-provided目录，其中的文件如图12.28所示。

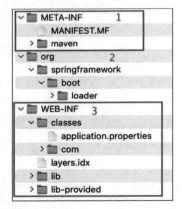

图12.27 WAR 包的目录结构

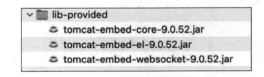

图12.28 lib-provided 目录下的文件

图12.28所示的这些JAR包正是在pom.xml中通过spring-boot-starter-tomcat配置的依赖，而spring-boot-starter-tomcat的scope为provided。

12.3.2 运行项目

Spring Boot项目打包生成的无论是JAR包还是WAR包，都是可执行的，也就是可以通过java -jar命令来运行。这种设计使得我们部署项目变得十分方便。

第 12 章　Spring Boot 项目的测试和部署

【示例12.8】　直接运行Spring Boot项目的WAR包

上面的示例中，项目demowar的程序启动类DemowarApplication中添加过一条日志输出，这里直接使用java -jar命令运行demowar项目的WAR包，并观察日志输出来验证WAR包被运行。

在IDEA的Terminal面板中，通过命令行进入demowar项目的路径，然后执行如下命令：

```
java -jar target/demowar-0.0.1-SNAPSHOT.war
```

此时，Terminal面板中的输出如图12.29所示。可以看到在前面的示例中添加的日志输出，说明WAR包被成功执行了。

```
2021-08-27 22:16:14.787  INFO 3545 --- [           main] com.example.demowar.De
mowarApplication         : Started DemowarApplication in 4.609 seconds (JVM running f
or 5.492)
demowar 启动成功！
```

图 12.29　项目启动日志输出

JAR包和WAR包可以执行，主要依赖于spring-boot-loader所提供的功能。在查看JAR包目录结构的示例中，看到过MANIFEST.MF文件中定义的Main-Class属性，其值为org.springframework.boot.loader.JarLauncher，这正是spring-boot-loader中提供的Launcher之一。如果是WAR包，那么Main-Class就是org.springframework.boot.loader.WarLauncher，用来解析和加载WAR包结构的路径。